Putting **Essential Understanding** of

Multiplication and Division into Practice

in Grades
3–5

John Lannin
University of Missouri
Columbia, Missouri

Kathryn Chval
University of Missouri
Columbia, Missouri

Dusty Jones
Sam Houston State University
Huntsville, Texas

Kathryn Chval
Volume Editor
University of Missouri
Columbia, Missouri

Barbara J. Dougherty
Series Editor
University of Missouri
Columbia, Missouri

NATIONAL COUNCIL OF
TEACHERS OF MATHEMATICS

www.nctm.org/more4u
Access code: MDV14347

Copyright © 2013 by
The National Council of Teachers of Mathematics, Inc.
1906 Association Drive, Reston, VA 20191-1502
(703) 620-9840; (800) 235-7566; www.nctm.org
All rights reserved

Library of Congress Cataloging-in-Publication Data

Lannin, John, author.
  Putting essential understanding of multiplication and division into practice in grades
3-5 / John Lannin, University of Missouri, Columbia, Missouri, Kathryn Chval, University
of Missouri, Columbia, Missouri, Dusty Jones, Sam Houston State University, Huntsville,
Texas; Barbara J. Dougherty, series editor, University of Missouri, Columbia, Missouri.
      pages cm.
  Includes bibliographical references.
  ISBN 978-0-87353-715-5
  1. Multiplication—Study and teaching (Middle school) 2. Division—Study and teaching
(Middle school) 3. Problem solving—Study and teaching (Middle school) I. Chval, Kathryn B.
(Kathryn Bouchard), author, editor. II. Jones, Dusty, author. III. Title.
  QA115.L347 2013
  372.7'2049—dc23
                        2013029108

The National Council of Teachers of Mathematics is the public voice of mathematics
education, providing vision, leadership, and professional development to support teachers
in ensuring equitable mathematics learning of the highest quality for all students.

When forms, problems, or sample documents are included or are made available
on NCTM's website, their use is authorized for educational purposes by educators
and noncommercial or nonprofit entities that have purchased this book. Except
for that use, permission to photocopy or use material electronically from *Putting
Essential Understanding of Multiplication and Division into Practice in Grades 3–5*
must be obtained from www.copyright.com or by contacting Copyright Clear-
ance Center, Inc. (CCC), 222 Rosewood Drive, Danvers, MA 01923, 978-750-8400.
CCC is a not-for-profit organization that provides licenses and registration for a
variety of users. Permission does not automatically extend to any items identified
as reprinted by permission of other publishers or copyright holders. Such items
must be excluded unless separate permissions are obtained. It is the responsibility
of the user to identify such materials and obtain the permissions.

The publications of the National Council of Teachers of Mathematics present a
variety of viewpoints. The views expressed or implied in this publication, unless
otherwise noted, should not be interpreted as official positions of the Council.

Printed in the United States of America

Contents

Chapter 3
The Meaning of Division

Chapter 4
Multiplication and Division Properties

Accompanying Materials at More4U

Foreword

Teaching mathematics in prekindergarten–grade 12 requires knowledge of mathematical content and developmentally appropriate pedagogical knowledge to provide students with experiences that help them learn mathematics with understanding, while they reason about and make sense of the ideas that they encounter.

In 2010 the National Council of Teachers of Mathematics (NCTM) published the first book in the Essential Understanding Series, focusing on topics that are critical to the mathematical development of students but often difficult to teach. Written to deepen teachers' understanding of key mathematical ideas and to examine those ideas in multiple ways, the Essential Understanding Series was designed to fill in gaps and extend teachers' understanding by providing a detailed survey of the big ideas and the essential understandings related to particular topics in mathematics.

The Putting Essential Understanding into Practice Series builds on the Essential Understanding Series by extending the focus to classroom practice. These books center on the pedagogical knowledge that teachers must have to help students master the big ideas and essential understandings at developmentally appropriate levels.

To help students develop deeper understanding, teachers must have skills that go beyond knowledge of content. The authors demonstrate that for teachers—

- understanding student misconceptions is critical and helps in planning instruction;

- knowing the mathematical content is not enough—understanding student learning and knowing different ways of teaching a topic are indispensable;

- constructing a task is important because the way in which a task is constructed can aid in mediating or negotiating student misconceptions by providing opportunities to identify those misconceptions and determine how to address them.

Through detailed analysis of samples of student work, emphasis on the need to understand student thinking, suggestions for follow-up tasks with the potential to move students forward, and ideas for assessment, the Putting Essential Understanding into Practice Series demonstrates best practice for developing students' understanding of mathematics.

The ideas and understandings that the Putting Essential Understanding into Practice Series highlights for student mastery are also embodied in the Common Core State

Standards for Mathematics, and connections with these new standards are noted throughout each book.

On behalf of the Board of Directors of NCTM, I offer sincere thanks to everyone who has helped to make this new series possible. Special thanks go to Barbara J. Dougherty for her leadership as series editor and to all the authors for their work on the Putting Essential Understanding into Practice Series. I join the project team in welcoming you to this special series and extending best wishes for your ongoing enjoyment—and for the continuing benefits for you and your students—as you explore Putting Essential Understanding into Practice!

Linda M. Gojak
President, 2012–2014
National Council of Teachers of Mathematics

Preface

The Putting Essential Understanding into Practice Series explores the teaching of mathematics topics in grades K–12 that are difficult to learn and to teach. Each volume in this series focuses on specific content from one volume in NCTM's Essential Understanding Series and links it to ways in which those ideas can be taught successfully in the classroom.

Thus, this series builds on the earlier series, which aimed to present the mathematics that teachers need to know and understand well to teach challenging topics successfully to their students. Each of the earlier books identified and examined the big ideas related to the topic, as well as the "essential understandings"—the associated smaller, and often more concrete, concepts that compose each big idea.

Taking the next step, the Putting Essential Understanding into Practice Series shifts the focus to the specialized pedagogical knowledge that teachers need to teach those big ideas and essential understandings effectively in their classrooms. The Introduction to each volume details the nature of the complex, substantive knowledge that is the focus of these books—*pedagogical content knowledge*. For the topics explored in these books, this knowledge is both student centered and focused on teaching mathematics through problem solving.

Each book then puts big ideas and essential understandings related to the topic under a high-powered teaching lens, showing in fine detail how they might be presented, developed, and assessed in the classroom. Specific tasks, classroom vignettes, and samples of student work serve to illustrate possible ways of introducing students to the ideas in ways that will enable students not only to make sense of them now but also to build on them in the future. Items for readers' reflection appear throughout and offer teachers additional opportunities for professional development.

The final chapter of each book looks at earlier and later instruction on the topic. A look back highlights effective teaching that lays the earlier foundations that students are expected to bring to the current grades, where they solidify and build on previous learning. A look ahead reveals how high-quality teaching can expand students' understanding when they move to more advanced levels.

Each volume in the Putting Essential Understanding into Practice Series also includes appendixes that list the big ideas and essential understandings related to the topic, detail resources for teachers, and present the tasks discussed in the book. These materials, which are available to readers both in the book and online at www.nctm.org/more4u, are intended to extend and enrich readers' experiences and possibilities for using the book. Readers can gain online access to these materials by going to the More4U website and entering the code that appears on the book's title page. They can then print out these materials for personal or classroom use.

Because the topics chosen for both the earlier Essential Understanding Series and this successor series represent areas of mathematics that are widely regarded as challenging to teach and to learn, we believe that these books fill a tangible need for teachers. We hope that as you move through the tasks and consider the associated classroom implementations, you will find a variety of ideas to support your teaching and your students' learning.

Acknowledgments from the Authors

We would like to thank the administrators and teachers at Paxton Keeley Elementary School and P. K. Yonge Developmental Research School for collaborating with us on the material for this volume. We would also like to thank their third-, fourth-, and fifth-grade students who shared their mathematical thinking with us. In addition, we extend thanks to Chris Bowling for his assistance in creating figures and scanning students' work, Anita Draper for her thoughtful copyediting of this volume, and Rachel Newman and Lina Trigos for supporting our efforts.

Finally, Dusty Jones would like to thank the University of Florida, where he served as a visiting professor while writing this book.

Introduction

Shulman (1986, 1987) identified seven knowledge bases that influence teaching:

1. Content knowledge

2. General pedagogical knowledge

3. Curriculum knowledge

4. Knowledge of learners and their characteristics

5. Knowledge of educational contexts

6. Knowledge of educational ends, purposes, and values

7. Pedagogical content knowledge

The specialized content knowledge that you use to transform your understanding of mathematics content into ways of teaching is what Shulman identified as item 7 on this list—*pedagogical content knowledge* (Shulman 1986). This is the knowledge that is the focus of this book—and all the volumes in the Putting Essential Understanding into Practice Series.

Pedagogical Content Knowledge

In mathematics teaching, pedagogical content knowledge includes at least four indispensable components:

1. Knowledge of curriculum for mathematics

2. Knowledge of assessments for mathematics

3. Knowledge of instructional strategies for mathematics

4. Knowledge of student understanding of mathematics (Magnusson, Krajcik, and Borko 1999)

These four components are linked in significant ways to the content that you teach.

Even though it is important for you to consider how to structure lessons, deciding what group and class management techniques you will use, how you will allocate time, and what will be the general flow of the lesson, Shulman (1986) noted that it is even more important to consider *what* is taught and the *way* in which it is taught. Every day, you make at least five essential decisions as you determine—

1. which explanations to offer (or not);

2. which representations of the mathematics to use;

3. what types of questions to ask;

4. what depth to expect in responses from students to the questions posed; and

5. how to deal with students' misunderstandings when these become evident in their responses.

Your pedagogical content knowledge is the unique blending of your content expertise and your skill in pedagogy to create a knowledge base that allows you to make robust instructional decisions. Shulman (1986, p. 9) defined pedagogical content knowledge as "a second kind of content knowledge…, which goes beyond knowledge of the subject matter per se to the dimension of subject matter knowledge *for teaching*." He explained further:

> Pedagogical content knowledge also includes an understanding of what makes the learning of specific topics easy or difficult: the conceptions and preconceptions that students of different ages and backgrounds bring with them to the learning of those most frequently taught topics and lessons. (p. 9)

If you consider the five decision areas identified at the top of the page, you will note that each of these requires knowledge of the mathematical content and the associated pedagogy. For example, an explanation that you give your students about multiplication requires that you take into account their knowledge of models of multiplication and important ideas that you want students to understand about the effect of multiplication on the numbers involved. Your knowledge of multiplication and division can help you craft tasks and questions that provide counterexamples and ways to guide your students in seeing connections across multiple number systems. As you establish the content, complete with learning goals, you then need to consider how to move your students from their initial understandings to deeper ones, building rich connections along the way.

The instructional sequence that you design to meet student learning goals has to take into consideration the misconceptions and misunderstandings that you might expect to encounter (along with the strategies that you expect to use to negotiate them), your expectation of the level of difficulty of the topic for your students, the progression of experiences in which your students will engage, appropriate collections of representations for the content, and relationships between and among multiplication and division and other topics.

Model of Teacher Knowledge

Grossman (1990) extended Shulman's ideas to create a model of teacher knowledge with four domains (see fig. 0.1):

1. Subject-matter knowledge

2. General pedagogical knowledge

3. Pedagogical content knowledge

4. Knowledge of context

Subject-matter knowledge includes mathematical facts, concepts, rules, and relationships among concepts. Your understanding of the mathematics affects the way in which you teach the content—the ideas that you emphasize, the ones that you do not, particular algorithms that you use, and so on (Hill, Rowan, and Ball 2005).

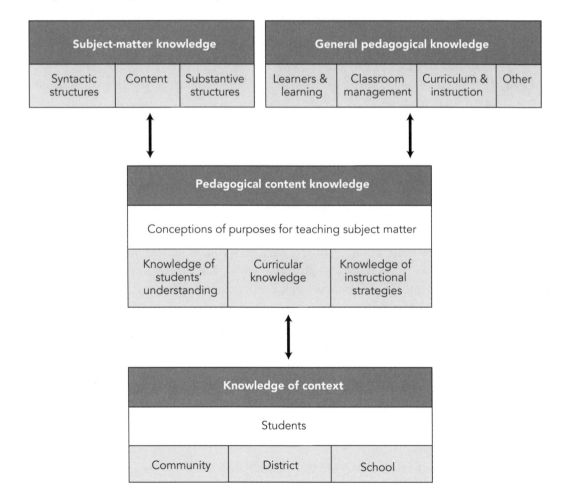

Fig. 0.1. Grossman's (1990, p. 5) model of teacher knowledge

Your pedagogical knowledge relates to the general knowledge, beliefs, and skills that you possess about instructional practices. These include specific instructional strategies that you use, the amount of wait time that you allow for students' responses to questions or tasks, classroom management techniques that you use for setting expectations and organizing students, and your grouping techniques, which might include having your students work individually or cooperatively or collaboratively, in groups or pairs. As Grossman's model indicates, your understanding and interpretation of the environment of your school, district, and community can also have an impact on the way in which you teach a topic.

Note that pedagogical content knowledge has four aspects, or components, in Grossman's (1990) model:

1. Conceptions of purposes for teaching

2. Knowledge of students' understanding

3. Knowledge of curriculum

4. Knowledge of instructional strategies

Each of these components has specific connections to the classroom. It is useful to consider each one in turn.

First, when you think about the goals that you want to establish for your instruction, you are focusing on your conceptions of the purposes for teaching. This is a broad category but an important one because the goals that you set will define learning outcomes for your students. These conceptions influence the other three components of pedagogical content knowledge. Hence, they appropriately occupy their overarching position in the model.

Second, your knowledge of your students' understanding of the mathematics content is central to good teaching. To know what your students understand, you must focus on both their conceptions and their misconceptions. As teachers, we all recognize that students develop naïve understandings that may or may not be immediately evident to us in their work or discourse. These can become deep-rooted misconceptions that are not simply errors that students make. Misconceptions may include incorrect generalizations that students have developed, such as thinking that multiplying is simply a shortcut for adding. These generalizations may even be predictable notions that students exhibit as part of a developmental trajectory, such as thinking that multiplying two numbers results in a product that is larger than either of the numbers being multiplied.

Part of your responsibility as a teacher is to present tasks or to ask questions that can bring misconceptions to the forefront. Once you become aware of misconceptions in students' thinking, you then have to determine the next instructional steps. The mathematical ideas presented in this volume focus on common misconceptions that students form in relation to specific topics—multiplication and division in grades 3–5. This book shows how the type of task selected and the sequencing of carefully developed questions can bring the misconceptions to light, as well as how particular teachers took the next instructional steps to challenge the students' misconceptions.

Third, curricular knowledge for mathematics includes multiple areas. Your teaching may be guided by a set of standards such as the Common Core State Standards for Mathematics (CCSSM; National Governors Association Center for Best Practices and Council of Chief State School Officers 2010) or other provincial, state, or local standards. You may in fact use these standards as the learning outcomes for your students. Your textbook is another source that may influence your instruction. With any textbook also comes a particular philosophical view of mathematics, mathematics teaching, and student learning. Your awareness and understanding of the curricular perspectives related to the choice of standards and the selection of a textbook can help to determine how you actually enact your curriculum. Moreover, your district or school may have a pacing guide that influences your delivery of the curriculum. In this book, we can focus only on the alignment of the topics presented with broader curricular perspectives, such as CCSSM. However, your own understanding of and expertise with your other curricular resources, coupled with the parameters defined by the expected student outcomes from standards documents, can provide the specificity that you need for your classroom.

In addition to your day-to-day instructional decisions, you make daily decisions about which tasks from curricular materials you can use without adaptation, which tasks you will need to adapt, and which tasks you will need to create on your own. Once you select or develop meaningful, high-quality tasks and use them in your mathematics lesson, you have launched what Yinger (1988) called "a three-way conversation between teacher, student, and problem" (p. 86). This process is not simple—it is complex because how students respond to the problem or task is directly linked to your next instructional move. That means that you have to plan multiple instructional paths to choose among as students respond to those tasks.

Knowledge of the curriculum goes beyond the curricular materials that you use. You also consider the mathematical knowledge that students bring with them from grade 2 and what they should learn by the end of grade 5. The way in which you teach a foundational concept or skill has an impact on the way in which students will interact with and learn later related content. For example, the types of representations that you include in your introduction of multiplication and division are the ones that your students will use to evaluate other representations and ideas in later grades.

Fourth, knowledge of instructional strategies is essential to pedagogical content knowledge. Having a wide array of instructional strategies for teaching mathematics is central to effective teaching and learning. Instructional strategies, along with knowledge of the curriculum, may include the selection of mathematical tasks, together with the way in which those tasks will be enacted in the classroom. Instructional strategies may also include the way in which the mathematical content will be structured for students. You may have very specific ways of thinking about how you will structure your presentation of a mathematical idea—not only how you will sequence the introduction and development of the idea, but also how you will present that idea to your students. Which examples should you select, and which questions should you ask? What representations should you use? Your knowledge of instructional strategies, coupled with your knowledge of your curriculum, permits you to align the selected mathematical tasks closely with the way in which your students perform those tasks in your classroom.

The instructional approach in this volume combines a student-centered perspective with an approach to mathematics through problem solving. A student-centered approach is characterized by a shared focus on student and teacher conversations, including interactions among students. Students who learn through such an approach are active in the learning process and develop ways of evaluating their own work and one another's in concert with the teacher's evaluation.

Teaching through problem solving makes tasks or problems the core of mathematics teaching and learning. The introduction to a new topic consists of a task that students work through, drawing on their previous knowledge while connecting it with new ideas. After students have explored the introductory task (or tasks), their consideration of solution methods, the uniqueness or multiplicity of solutions, and extensions of the task create rich opportunities for discussion and the development of specific mathematical concepts and skills.

By combining the two approaches, teachers create a dynamic, interactive, and engaging classroom environment for their students. This type of environment promotes the ability of students to demonstrate CCSSM's Standards for Mathematical Practice while learning the mathematics at a deep level.

The chapters that follow will show that instructional sequences embed all the characteristics of knowledge of instructional strategies that Grossman (1990) identifies. One component that is not explicit in Grossman's model but is included in a model developed by Magnusson, Krajcik, and Borko (1999) is the knowledge of assessment. Your knowledge of assessment in mathematics plays an important role in guiding your instructional decision-making process.

There are different types of assessments, each of which can influence the evidence that you collect as well as your view of what students know (or don't know) and how they know what they do. Your interpretation of what students know is also related to your view of what constitutes "knowing" in mathematics. As you examine the tasks, classroom vignettes, and samples of student work in this volume, you will notice that teacher questioning permits formative assessment that supplies information that spans both conceptual and procedural aspects of understanding. *Formative assessment*, as this book uses the term, refers to an appraisal that occurs during an instructional segment, with the aim of adjusting instruction to meet the needs of students more effectively (Popham 2006). Formative assessment does not always require a paper-and-pencil product but may include questions that you ask or tasks that students complete during class.

The information that you gain from student responses can provide you with feedback that guides the instructional flow, while giving you a sense of how deeply (or superficially) your students understand a particular idea—or whether they hold a misconception that is blocking their progress. As you monitor your students' development of rich understanding, you can continually compare their responses with your expectations and then adapt your instructional plans to accommodate their current levels of development. Wiliam (2007, p. 1054) described this interaction between teacher expectations and student performance in the following way:

> It is therefore about assessment functioning as a bridge between teaching and learning, helping teachers collect evidence about student achievement in order to adjust instruction to better meet student learning needs, in real time.

Wiliam notes that for teachers to get the best information about student understandings, they have to know how to facilitate substantive class discussions, choose tasks that include opportunities for students to demonstrate their learning, and employ robust and effective questioning strategies. From these strategies, you must then interpret student responses and scaffold their learning to help them progress to more complex ideas.

Characteristics of Tasks

The type of task that is presented to students is very important. Tasks that focus only on procedural aspects may not help students learn a mathematical idea deeply.

Superficial learning may result in students forgetting easily, requiring reteaching, and potentially affecting how they understand mathematical ideas that they encounter in the future. Thus, the tasks selected for inclusion in this volume emphasize deep learning of significant mathematical ideas. These rich, "high-quality" tasks have the power to create a foundation for more sophisticated ideas and support an understanding that goes beyond "how" to "why." Figure 0.2 identifies the characteristics of a high-quality task.

As you move through this volume, you will notice that it sequences tasks for each mathematical idea so that they provide a cohesive and connected approach to the identified concept. The tasks build on one another to ensure that each student's thinking becomes increasingly sophisticated, progressing from a novice's view of the content to a perspective that is closer to that of an expert. We hope that you will find the tasks useful in your own classes.

A high-quality task has the following characteristics:
Aligns with relevant mathematics content standard(s)
Encourages the use of multiple representations
Provides opportunities for students to develop and demonstrate the mathematical practices
Involves students in an inquiry-oriented or exploratory approach
Allows entry to the mathematics at a low level (all students can begin the task) but also has a high ceiling (some students can extend the activity to higher-level activities)
Connects previous knowledge to new learning
Allows for multiple solution approaches and strategies
Engages students in explaining the meaning of the result
Includes a relevant and interesting context

Fig. 0.2. Characteristics of a high-quality task

Types of Questions

The questions that you pose to your students in conjunction with a high-quality task may at times cause them to confront ideas that are at variance with or directly contradictory to their own beliefs. The state of mind that students then find themselves in is called *cognitive dissonance*, which is not a comfortable state for students—or, on occasion, for the teacher. The tasks in this book are structured in a way that forces students to deal with two conflicting ideas. However, it is through the process of negotiating the contradictions that students come to know the content much more deeply. How the teacher handles this negotiation determines student learning.

You can pose three types of questions to support your students' process of working with and sorting out conflicting ideas. These questions are characterized by their potential to encourage reversibility, flexibility, and generalization in students' thinking (Dougherty 2001). All three types of questions require more than a one-word or one-number answer. Reversibility questions are those that have the capacity to change the direction of students' thinking. They often give students the solution and require them to create the corresponding problem. A flexibility question can be one of two types: it can ask students to solve a problem in more than one way, or it can ask them to compare and contrast two or more problems or determine the relationship between or among concepts and skills. Generalization questions also come in two types: they ask students to look at multiple examples or cases and find a pattern or make observations, or they ask them to create a specific example of a rule, conjecture, or pattern. Figure 0.3 provides examples of reversibility, flexibility, and generalization questions related to multiplication and division.

Type of question	Example
Reversibility question	Find two numbers whose product is 36.
Flexibility question	How are multiplication and division alike? How are they different?
Flexibility question	Show two ways to multiply 36×10.
Generalization question	Find two numbers that when multiplied give a product that is an odd number. What do you notice about the two factors?
Generalization question	$2 \times 7 = ?$ $3 \times 7 = ?$ $4 \times 7 = ?$ $5 \times 7 = ?$ What patterns do you notice?

Fig. 0.3. Examples of reversibility, flexibility, and generalization questions

Conclusion

The Introduction has provided a brief overview of the nature of—and necessity for—pedagogical content knowledge. This knowledge, which you use in your classroom every day, is the indispensable medium through which you transmit your understanding of the big ideas of the mathematics to your students. It determines your selection of appropriate, high-quality tasks and enables you to ask the types of questions that will not only move your students forward in their understanding but also allow you to determine the depth of that understanding.

The chapters that follow describe important ideas related to learners, curricular goals, instructional strategies, and assessment that can assist you in transforming your students' knowledge into formal mathematical ideas related to multiplication and division. These chapters provide specific examples of mathematical tasks and student thinking for you to analyze to develop your pedagogical content knowledge for teaching multiplication and division in grades 3–5 or to give you ideas to help other colleagues develop this knowledge. You will also see how to bring together and interweave your knowledge of learners, curriculum, instructional strategies, and assessment to support your students in grasping the big ideas and essential understandings and using them to build more sophisticated knowledge.

Students in grades 3–5 have had many experiences in earlier grades and outside the classroom that have shaped their initial understanding of multiplication and division. They frequently demonstrate an understanding of multiplication and division in a particular context or in connection with a specific model or tool. Yet, in other situations, these same students do not demonstrate the same understanding. As their teacher, you must understand the meaning that your students have begun to give to these operations so that you can extend this knowledge, build on it, and see whether or how it differs from the formal mathematical knowledge that they need to be successful in reasoning about and using multiplication and division. You have the important responsibility of assessing their current knowledge related to the big ideas of multiplication and division as well as their understanding of various representations and their power and limitations. Your understanding will facilitate and reinforce your instructional decisions. Teaching the big ideas and essential understandings related to multiplication and division is obviously a very challenging and complex task.

Chapter 1
The Meaning of Multiplication

Essential Understanding 1*a*
In the multiplicative expression $A \times B$, A can be defined as a *scaling factor*.
Essential Understanding 1*c*
A situation that can be represented by multiplication has an element that represents the scalar and an element that represents the quantity to which the scalar applies.
Essential Understanding 1*d*
A scalar definition of multiplication is useful in representing and solving problems beyond whole number multiplication and division.

Multiplication is a scalar process involving two quantities, with one quantity–the *multiplier*–serving as a scaling factor and specifying how the operation resizes, or rescales, the other quantity–the *multiplicative unit*. The rescaled result is the *product* of the multiplication. Understanding multiplication as a scalar operation on whole numbers as well as other numbers is the foundation of multiplicative thinking and underlies Essential Understandings 1*a*, 1*c*, and 1*d*, presented in *Developing Essential Understanding of Multiplication and Division for Teaching Mathematics in Grades 3–5* (Otto et al. 2011). This chapter focuses on helping students recognize situations that call for multiplicative thinking and assessing their understanding of the elements that make these situations multiplicative.

Working toward Essential Understandings 1*a*, 1*c*, and 1*d*

As Otto and colleagues (2011) discuss, "Multiplication is a fundamental operation that is used to solve everyday problems" (p. 10). Yet, many students struggle to develop a deep understanding of multiplication and the underlying ideas related to it. To help students develop such understanding, teachers need to design, adapt, or select worthwhile mathematical tasks for them to work with, interpret the responses that they give, and make instructional decisions on the basis of the thinking that they reveal. These critical practices require specialized knowledge. To begin exploring students' development of this understanding, analyze the problems in figure 1.1. As you consider these problems, think about the questions posed in Reflect 1.1.

Reflect 1.1

Which of the problems shown in figure 1.1 call for multiplicative reasoning? Which call for additive reasoning?

How do these problems compare with the problems that you use to help your students develop additive and multiplicative reasoning?

What are the benefits of using tasks that require students to compare and contrast situations involving multiplicative and additive reasoning?

1. Phil ran 2 miles. Sally ran 3 times the distance that Phil ran. How many miles did Sally run?

2. Phil ran 2 miles. Sally ran 3 more miles than Phil. How many miles did Sally run?

3. Phil ran $3/4$ of a mile. Sally ran $2/3$ of the distance that Phil ran. How many miles did Sally run?

4. Phil ran $3/4$ of a mile. Sally ran $2/3$ of a mile more than Phil ran. How many miles did Sally run?

Fig. 1.1. Contextual problems about the number of miles that Sally ran: additive or multiplicative situations?

The four situations in figure 1.1 have some similarities. For example, they all pose the same question: "How many miles did Sally run?" Yet, students' responses to subtle differences in the situations provide opportunities to assess their multiplicative reasoning. Problems 2 and 4 require students to add the two values to determine the total number of miles that Sally ran. By contrast, problems 1 and 3 require students to think multiplicatively. Draw diagrams to represent problems 1 and 3, and then respond to the questions in Reflect 1.2.

Reflect 1.2

What does reasoning multiplicatively mean?

What does reasoning additively mean?

Clearly, you want your students to do more than just provide answers to multiplication facts—telling you only, for example, that 3 × 5 is 15. Students need to build an understanding of the meaning of multiplicative situations and learn to reason multiplicatively. These essential understandings and competencies will support their future learning when they encounter topics such as prime and composite numbers, factorization and prime factorization, factor and greatest common factor, multiple, area and volume, proportional reasoning, mean, algebraic expressions, linear functions, and place value (Otto et al. 2011).

One key aspect of your students' understanding that you, as their teacher, need to encourage is the development of multiplicative reasoning that extends beyond a view of multiplication as repeated addition (Jacob and Willis 2001). Students should relate multiplicative reasoning to iterating—that is, to making multiple copies—and partitioning sets of objects as well as to the length, area, and volume of physical quantities. When students see the mathematical expression 3 × 5, for example, they should be able to view 5 as the multiplicative unit (also called the *multiplicand*) and 3 as the scaling factor (also called the *multiplier*) for that multiplicative unit. Such a perspective not only involves recognizing the multiplicative unit, but also being able to iterate it—make multiple copies of it. The expression 3 × 5, for example, means 3 copies of 5 or 3 groups of 5. This interpretation of 3 × 5 reflects a critical meaning of multiplication that students must establish.

Students who reason multiplicatively can "see" a multiplicative unit and create multiple copies of it. An initial view of multiplication should involve understanding what it means to create 1, 2, 3, 5, or more copies of a given unit. Note that some researchers refer to the multiplicative unit as the *composite unit* (e.g., Steffe 1992; Lamon 1994; Tzur et al. 2013).

Eventually, we want students to recognize what it means to create $1/2$ of the multiplicative unit $2\frac{1}{2}$, obtaining the product $5/4$. Students who have a deeper understanding of multiplicative reasoning will begin to make multiplicative comparisons and express them in statements such as, "This is half as much as I had before."

Jacob and Willis (2001, p. 307) emphasized the importance of three aspects of multiplicative situations:

It was the work of Kouba (1989), Steffe (1992), Mulligan and Mitchelmore (1997) and Mulligan and Watson (1998) that led to the conclusion that children must first come to recognise multiplicative situations as involving three aspects: groups of equal size (a multiplicand), numbers of groups (the multiplier), and a total amount (the product). When they can construct and coordinate these factors in both multiplication and division problems prior to carrying out the count, they are thinking multiplicatively.

Let's consider these three critical aspects—the multiplicand (or multiplicative unit); the multiplier; and the product—in the case of two problem situations:

1. Elizabeth has 3 bags with 5 apples in each bag. How many apples does Elizabeth have?

2. Elizabeth has 3 pieces of ribbon. Each ribbon is 5 inches long. What is the total length of the ribbon that Elizabeth has?

Both of these situations can be expressed as $3 \times 5 = 15$. Depending on the context, students approaching these problems work with objects or units of measure—apples in problem 1, inches in problem 2. They must conceptualize some number of objects or units of measure as the multiplicative unit—5 apples in problem 1, 5 inches of ribbon in problem 2. This number (or quantity) of objects or units of measure is the *multiplicand*—in both of these cases, 5. Figure 1.2 illustrates these multiplicative units of 5 apples and 5 inches of ribbon, configured in problem 2 as a 5-inch segment of ribbon.

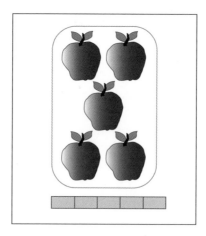

Fig. 1.2. The multiplicative units for the apple and ribbon situations in problems 1 and 2

Students then must iterate the multiplicative unit a number of times. The *multiplier*, or scalar factor, represents the number of iterations. In problems 1 and 2, both of which present situations that can be expressed as 3 × 5, the multiplier, or scalar factor, is 3, as figure 1.3 illustrates. The total number of objects or units of measure after the operation of the scalar factor is the *product*, or 15, as illustrated in figure 1.4.

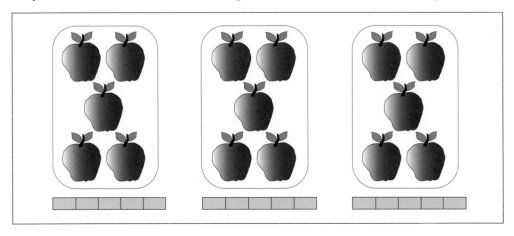

Fig. 1.3. The multiplier, 3, for the apple and ribbon situations in problems 1 and 2

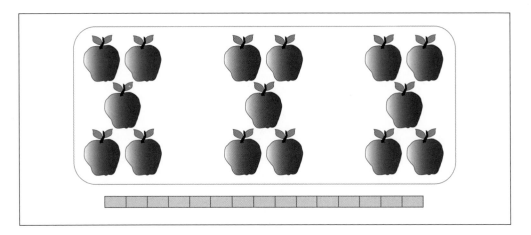

Fig. 1.4. The product

Consider another case, $^2/_3 × ^3/_4$, which extends the process of multiplication beyond whole numbers. Figure 1.5 presents a number line representation of $^2/_3 × ^3/_4$—a symbolic expression that could represent problem 3 in figure 1.1:

> Phil ran $^3/_4$ of a mile. Sally ran $^2/_3$ of the distance that Phil ran. How many miles did Sally run?

The multiplicand, or multiplicative unit, is $^3/_4$. Therefore, students using a number line representation need to understand that $^3/_4$ of a unit on the number line represents this multiplicative unit. The multiplier (scalar factor) is $^2/_3$, which means that students need to partition the multiplicative unit into 3 equal parts and iterate that part twice. The product is 2 copies of $^1/_3$ of a length of $^3/_4$, or $^2/_3$ of $^3/_4$, which is $^2/_4$ of a unit on the number line.

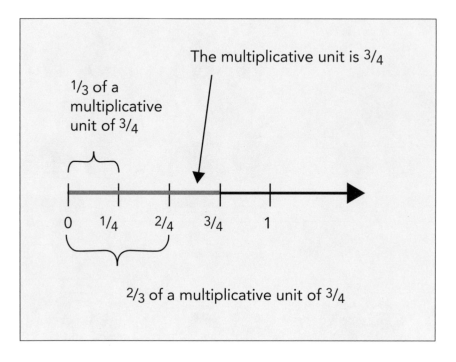

Fig 1.5. Illustrating $^2/_3 \times ^3/_4$ on a number line

Initially, students may have little sense of what a multiplicative unit is or that they can quantify a given situation in at least two different ways—one, by counting the number of objects within a multiplicative unit and the number of copies of the multiplicative unit, and the other, by counting the total number of objects in all the copies. For example, students should recognize that a diagram showing 7 circles with 3 smaller circles in each one could be viewed as 7 groups of 3 objects or, at the same time, as 21 objects. They should be able to identify the groups and the objects. However, such a dual view of the number of objects and the number of groups can be difficult for some children to coordinate.

To gain further insight into the different ways that students may view the multiplicative unit and the meaning of multiplication, consider the understanding and misunderstanding that third and fourth graders demonstrated in the work shown in figures 1.6–1.11. The third graders made representations of 7 groups of 3 and 7 × 3 and gave the product of 7 × 3, and the fourth graders created representations of 7 × 3. As you review the samples of student work, consider the questions in Reflect 1.3.

Reflect 1.3

How would you characterize the mathematical understandings and misunderstandings that the third- and fourth-grade students exhibit in their work in figures 1.6–1.11?

What specific strategies would you use, or questions would you ask, to move these students forward?

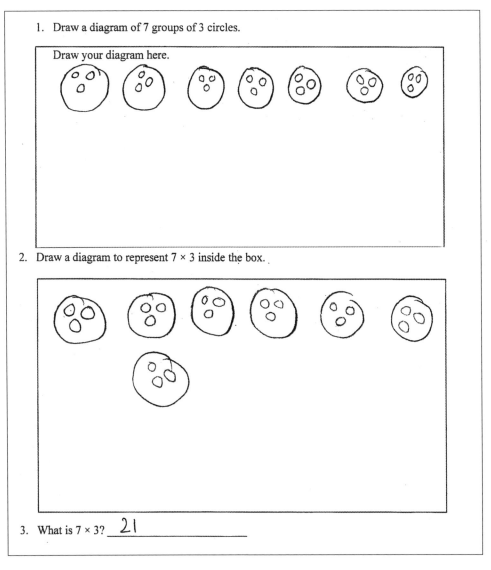

1. Draw a diagram of 7 groups of 3 circles.

Draw your diagram here.

2. Draw a diagram to represent 7 × 3 inside the box.

3. What is 7 × 3? ___21___

Fig 1.6. Alicia's (grade 3) representations of 7 groups of 3 and 7 × 3 and her response to 7 × 3 = ?

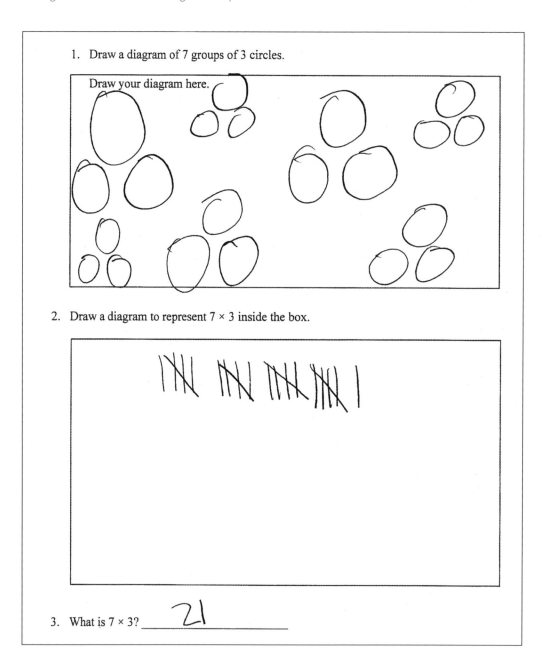

1. Draw a diagram of 7 groups of 3 circles.

   Draw your diagram here.

2. Draw a diagram to represent 7 × 3 inside the box.

3. What is 7 × 3? _____21_____

Fig 1.7. Phil's (grade 3) representations of 7 groups of 3 and 7 × 3 and his response to 7 × 3 = ?

1. Draw a diagram of 7 groups of 3 circles.

Draw your diagram here.

2. Draw a diagram to represent 7 × 3 inside the box.

I don't know

3. What is 7 × 3? __21__

Fig 1.8. Rachel's (grade 3) representations of 7 groups of 3 and 7 × 3 and her response to 7 × 3 = ?

1. Draw a diagram of 7 groups of 3 circles.

Draw your diagram here.

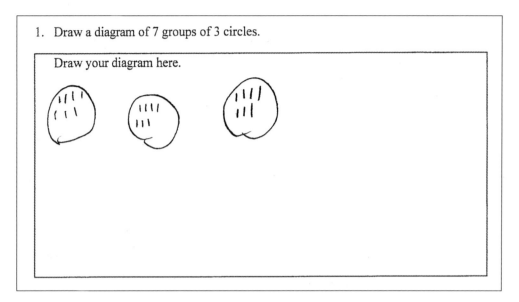

Fig 1.9. Kristen's (grade 3) representation of 7 groups of 3 circles

1. Draw a diagram to represent 7 × 3 inside the box.

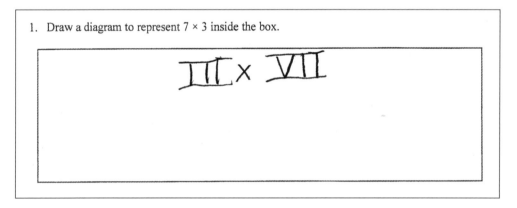

Fig 1.10. Aaron's (grade 4) representation of 7 × 3

1. Draw a diagram to represent 7 × 3 inside the box.

Fig 1.11. Jane's (grade 4) representation of 7 × 3

The work by Alicia, shown in figure 1.6, suggests that she connects the idea of 7 groups of 3 with the idea of 7 × 3 and also with the product of 7 × 3. She properly created 7 groups with 3 objects in each group, or multiplicative unit. The majority of 221 students in grades 3, 4, and 5 who completed these tasks created correct diagrams for 7 groups of 3 and 7 × 3. Yet, in some cases students struggled to create appropriate diagrams for these situations.

In the work displayed in figure 1.7, Phil demonstrates that he has established a meaning for 7 groups of 3 and that he knows that 7 × 3 is 21. However, whether he has connected the meaning of 7 groups of 3 with the meaning of 7 × 3 is unclear, since he simply drew 21 tally marks without creating groups of 3. Rachel, whose work appears in figure 1.8, computed 7 × 3 correctly but was unable to represent 7 × 3 or draw an appropriate diagram for 7 groups of 3. Kristen's work in figure 1.9 is the last sample from a third grader. By drawing 3 groups of 7 tally marks rather than 7 groups of 3, Kristen demonstrates a common confusion, shared by about 15 percent of the third graders surveyed.

Some students in fourth grade failed to connect the meaning of 7 × 3 with 7 groups of 3, as illustrated by the sample work in figures 1.10 and 1.11. Both Aaron (see fig. 1.10) and Jane (see fig. 1.11) created representations that demonstrate potential confusion between the symbolic expression for 7 × 3 and the pictorial representation of 7 groups of 3. As teachers, we need to be careful about our use of symbolic and pictorial representations for multiplication and division, and avoid mixing symbolic representations (using "×" and "÷") with pictorial representations showing groups of objects. As the students' work in figures 1.6–1.11 illustrates, such representations may demonstrate a superficial understanding of the meaning of multiplication and a lack of multiplicative reasoning.

Making multiplicative units with objects or units of measure as well as viewing the number of and number within multiplicative units is an indispensable foundational step that supports multiplicative reasoning. Thompson and Saldanha (2003) write, "As described by Confrey (1994) and Steffe (1988, 1994), multiplication can be introduced to students by asking them to think about quantities and numbers in settings in which they need to envision a multiplicity of identical objects" (p. 104). As students develop their ability to recognize multiplicative situations and deepen their use of multiplicative reasoning, they will spontaneously apply multiplicative thinking and make multiplicative comparisons in appropriate situations. Students' responses to particular tasks can make their degree of understanding apparent. Examine the task shown in figure 1.12 while considering the question in Reflect 1.4.

Reflect 1.4

What are the different ways in which students might correctly answer the comparison question in figure 1.12?

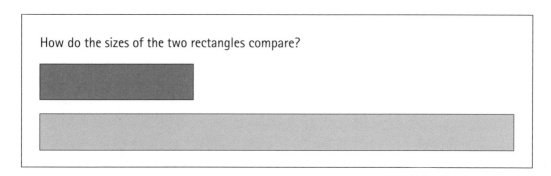

How do the sizes of the two rectangles compare?

Fig. 1.12. Rectangle Comparison task

Students with a beginning understanding of comparison may describe the gray rectangle as longer than the orange rectangle, without elaborating further. Other students at a slightly more advanced level may attempt to quantify or measure the amount by which the length of the gray rectangle exceeds the length of the orange rectangle. By describing a difference, these students demonstrate additive reasoning. Students who describe the length of one of the rectangles in terms of the length of the other—that is, by using one of the rectangles as a unit—are demonstrating multiplicative reasoning. Multiplicative reasoning would be evident in statements such as the following:

- The gray rectangle is more than twice as long as the orange rectangle.

- The gray rectangle is about three times as long as the orange rectangle.

- The orange rectangle is less than half as long as the gray rectangle.

- The orange rectangle is about one-third as long as the gray rectangle.

Thompson and Saldanha (2003, p. 104) discuss this important idea:

> We re-emphasize that when a curriculum starts with the idea that "___×___" means "some number of (or fraction of) some amount," it is not starting with the idea that "times" means to calculate. It is starting with the idea

that "times" means to envision something in a particular way—to think of copies (including parts of copies) of some amount. This is not to suggest that multiplication should not be about calculating. Rather, calculating is just one thing one might do when thinking of a product. Noncalculational ways to think of products will be important in comprehending situations in which multiplicative calculations might be useful.

The Common Core State Standards for Mathematics (CCSSM; National Governors Association Center for Best Practices and Council of Chief State School Officers 2010) emphasize the importance of distinguishing multiplicative and additive comparisons:

> Multiply or divide to solve word problems involving multiplicative comparison, e.g., by using drawings and equations with a symbol for the unknown number to represent the problem, distinguishing multiplicative comparison from additive comparison. (4.OA.A2, p. 29)

We designed the scenario shown in figure 1.13 to determine which third-grade students relied on additive reasoning to interpret and respond to a multiplicative situation. Reflect 1.5 provides guidance for considering students' thinking about the scenario in the figure.

Reflect 1.5

Describe ways in which students may correctly compare the numbers of star stickers that Morris and Leslie each used to make a picture of the night sky.

Which of these ways of comparing use multiplicative reasoning?

Which use additive reasoning?

Morris and Leslie both made pictures of the sky at night.
Morris used 60 star stickers for his picture.
Leslie used 20 star stickers for her picture.

Fig. 1.13. The Star Sticker scenario for third-grade students

We gave the Star Sticker scenario to 53 students in grade 3 and presented them with the reasoning of three fictitious students—Anna, Ben, and Cade:

- Anna said, "Morris used 3 times as many star stickers as Leslie."

- Ben said, "Leslie used 3 times as many star stickers as Morris."

- Cade said, "Leslie used 40 more stickers than Morris."

Of the three fictitious students, only Anna reasoned correctly. Both Anna and Ben used language related to multiplicative reasoning, though Ben's reasoning was incorrect. Cade, by contrast, came to an incorrect conclusion based on additive reasoning. Figures 1.14–1.16 show the responses of Corinne, Julia, and William to the reasoning of Anna, Ben, and Cade. Reflect 1.6 offers questions to guide your examination of the three third graders' responses.

Reflect 1.6

Figures 1.14–1.16 show the work of three third-grade students, Corinne, Julia, and William, in response to the reasoning of three fictitious students on the Star Sticker scenario, shown in figure 1.13.

Which third graders demonstrate multiplicative reasoning in their interpretations of the fictitious students' work?

What specific strategies would you use to move Corinne, Julia, and William forward?

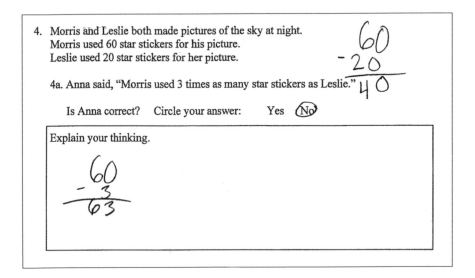

Fig. 1.14. Corinne's (grade 3) responses to the Star Sticker scenario

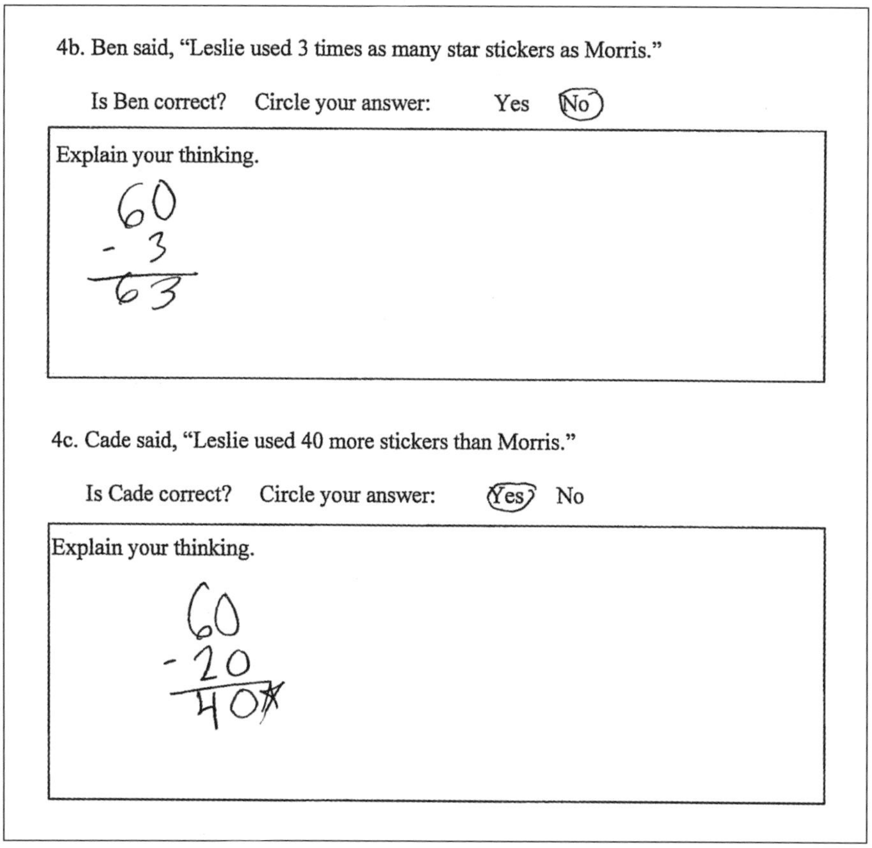

4b. Ben said, "Leslie used 3 times as many star stickers as Morris."

Is Ben correct?   Circle your answer: Yes   (No)

Explain your thinking.

$$\begin{array}{r} 60 \\ -\ 3 \\ \hline 63 \end{array}$$

4c. Cade said, "Leslie used 40 more stickers than Morris."

Is Cade correct?   Circle your answer: (Yes)   No

Explain your thinking.

$$\begin{array}{r} 60 \\ -\ 20 \\ \hline 40 \end{array}$$

Fig. 1.14. *Continued*

4. Morris and Leslie both made pictures of the sky at night.
   Morris used 60 star stickers for his picture.
   Leslie used 20 star stickers for her picture.

   4a. Anna said, "Morris used 3 times as many star stickers as Leslie."

   Is Anna correct?   Circle your answer: (Yes)   No

   Explain your thinking.

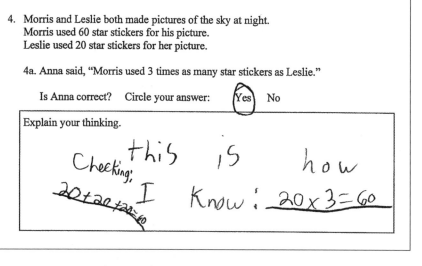

Checking: this is how

20+20+20=60   I know: 20 x 3 = 60

Fig. 1.15. Julia's (grade 3) response to the Star Sticker scenario

4b. Ben said, "Leslie used 3 times as many star stickers as Morris."

Is Ben correct?   Circle your answer: Yes   (No)

Explain your thinking.

I think this because I know that leslie has less Stars than less morris

4c. Cade said, "Leslie used 40 more stickers than Morris."

Is Cade correct?   Circle your answer: Yes   (No)

Explain your thinking.

my thinking:

$$\begin{array}{r} 60 \\ -20 \\ \hline 40 \end{array}$$ its not + 40   ro 20 ange

60

Fig. 1.15. *Continued*

4. Morris and Leslie both made pictures of the sky at night.
    Morris used 60 star stickers for his picture.
    Leslie used 20 star stickers for her picture.

4a. Anna said, "Morris used 3 times as many star stickers as Leslie."

Is Anna correct?   Circle your answer: (Yes)   No

Explain your thinking.   20 + 20 + 20 = 60

Fig. 1.16. William's (grade 3) response to the Star Sticker scenario

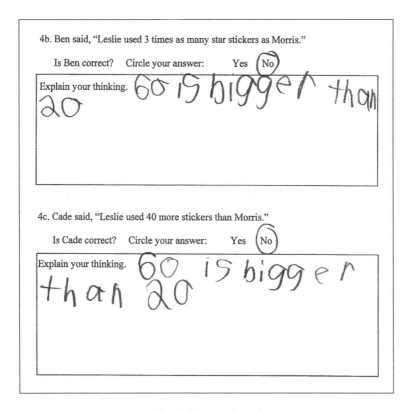

4b. Ben said, "Leslie used 3 times as many star stickers as Morris."

Is Ben correct?   Circle your answer: Yes   (No)

Explain your thinking.   60 is bigger than 20

4c. Cade said, "Leslie used 40 more stickers than Morris."

Is Cade correct?   Circle your answer: Yes   (No)

Explain your thinking.   60 is bigger than 20

Fig. 1.16. *Continued*

Figure 1.14 shows the work of Corinne, who appears to have used additive rather than multiplicative reasoning. Corinne rejected the multiplicative thinking of Anna and Ben, who compare the number of star stickers (20 and 60) used by Leslie and Morris by saying that one used "3 times as many" as the other. Concluding that this thinking was incorrect, Corinne offered the sum 60 + 3 = 63 in her explanation (although it is interesting to note that she wrote a subtraction symbol rather than a plus sign for addition). She also concluded that Cade was correct in saying that "Leslie used 40 more stickers than Morris," offering the difference 60 − 20 = 40 in her explanation and apparently not noticing that Cade's statement reversed Leslie and Morris; Morris in fact used more stickers than Leslie.

Julia, whose work is shown in figure 1.15, used multiplicative reasoning correctly and supported her agreement with Anna that "Morris used 3 times as many star stickers as Leslie" by writing both multiplication and addition statements. In disagreeing with Ben's assertion that "Leslie used 3 times as many star stickers as Morris," Julia pointed out that Leslie used fewer stickers than Morris.

Figure 1.16 shows the work of William, who appears to have iterated 3 groups of 20 in support of his acceptance of Anna's statement, "Morris used 3 times as many star stickers as Leslie." He offered identical explanations for his rejection of Ben's and Cade's thinking. In both cases, he wrote that "60 is bigger than 20," a comment echoed by many of his classmates. By itself, this observation, though sufficient to rebut Ben's and Cade's assertions, does not show evidence of either multiplicative reasoning or additive reasoning.

We share one final task that we used to assess students' understanding of the meaning of multiplication. Figure 1.17 shows a task that we gave to 44 third graders. This task presented students with a multiplicative situation and asked students to choose the number sentence that matched the situation. Examine the task and then respond to the question in Reflect 1.7.

Reflect 1.7

Figure 1.17 presents the Number Sentence task, set in the context of Marlene's muffins. Students must choose the number sentence that they could use to find the number of muffins that Marlene made, and they must explain their thinking.

What would you accept as evidence that a student understands the meaning of multiplication?

Marlene made 6 batches of muffins. There were 24 muffins in each batch. Which of the following number sentences could be used to find the number of muffins that she made?

Circle the correct number sentence below.

$6 \times$ ___ $= 24$

$6 + 24 =$ ___

$6 +$ ___ $= 24$

$6 \times 24 =$ ___

Explain your thinking.

Fig. 1.17. Number Sentence task

Figures 1.18–1.21 show responses from four third-grade students who completed the Number Sentence task. Examine their work, and then address the question posed in Reflect 1.8.

Reflect 1.8

Figures 1.18, 1.19, 1.20, and 1.21 show the responses of Tasha, Yakov, Mitchell, and Lois, respectively, to the Number Sentence task in figure 1.17. What conclusions would you draw about these third graders' understandings or misunderstandings of the meaning of multiplication?

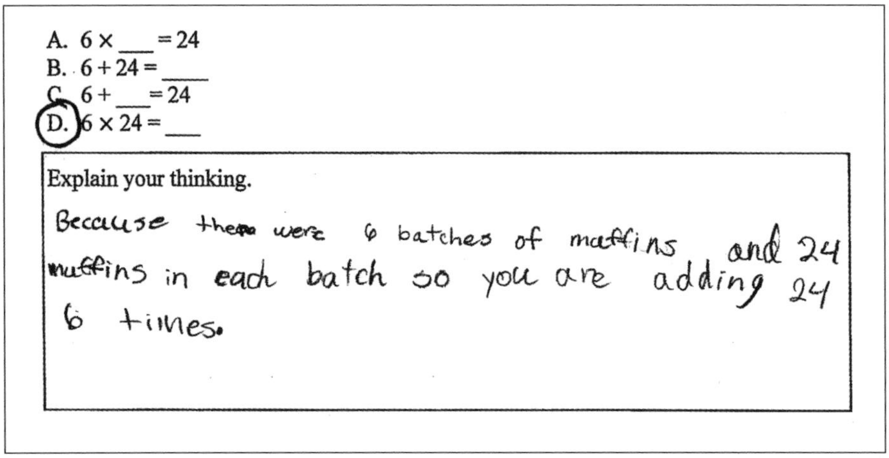

A. 6 × ___ = 24
B. 6 + 24 = ___
C. 6 + ___ = 24
D. 6 × 24 = ___ *(D is circled)*

Explain your thinking.

Because there were 6 batches of maffins and 24 muffins in each batch so you are adding 24 6 times.

Fig. 1.18. Tasha's (grade 3) response to the Number Sentence task

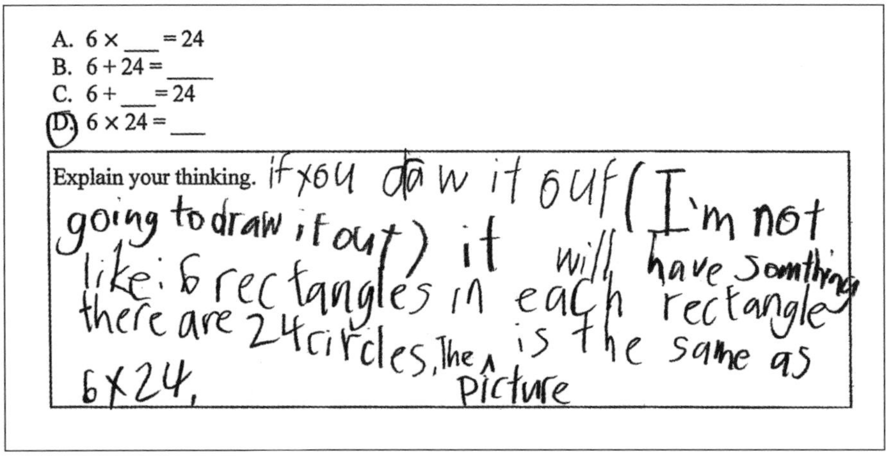

A. 6 × ___ = 24
B. 6 + 24 = ___
C. 6 + ___ = 24
D. 6 × 24 = ___ *(D is circled)*

Explain your thinking. if you daw it ouf (I'm not going to draw it out) it will have Somthing like: 6 rectangles in each rectangle there are 24 circles, The picture is the same as 6 × 24,

Fig. 1.19. Yakov's (grade 3) response to the Number Sentence task

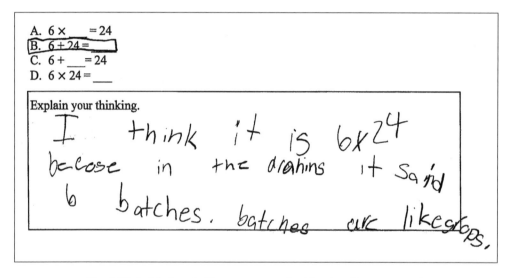

A. 6 × ___ = 24
B. 6 + 24 = ___
C. 6 + ___ = 24
D. 6 × 24 = 150

Explain your thinking.

I don't know

Fig. 1.20. Mitchell's (grade 3) response to the Number Sentence task

A. 6 × ___ = 24
B. 6 + 24 =
C. 6 + ___ = 24
D. 6 × 24 = ___

Explain your thinking.

I think it is 6×24 becose in the drahins it said 6 batches. batches are likegrops.

Fig. 1.21. Lois's (grade 3) response to the Number Sentence task

Tasha, Yakov, and Mitchell, whose work appears in figures 1.18, 1.19, and 1.20, respectively, all circled the correct number sentence, 6 × 24 = ___. Lois was the only one of the four students who did not, but her explanation, shown in figure 1.21, appears to address the correct sentence anyway.

Tasha explained her choice of 6 × 24 = ___ by describing the process of repeated addition: "so you are adding 24 6 times." Yakov described a picture that would represent 6 groups of 24. Although Mitchell marked the correct number sentence, he also filled in the blank with a number (150) that is not the product. Whether he gave 150 as an estimate or as the result of an error in his calculations is unclear because he did not explain his thinking. He gave no evidence of how he was thinking about the problem—did he perhaps reason correctly but was unable to communicate his reasoning on paper? Finally, Lois marked an incorrect statement but appears to have described a correct statement, and she supported her answer with the fact that batches are like groups.

In our sample of 44 grade 3 students, 32 (73 percent) answered this item correctly, including all four students whose work is discussed above. This same item (without the prompt for explanation) was provided to a nationally representative sample of fourth-grade students in the Sixth Mathematics Assessment of the National Assessment of Educational Progress (Kenney and Silver 1997), and 37 percent of students answered correctly. The most frequently selected incorrect answer was $6 \times \underline{\quad} = 24$, given as choice A (27 percent). In our third-grade sample, only 3 students (7 percent) selected choice A.

Even though we did not ask students to calculate the total number of muffins, 9 of the 32 students who marked the correct sentence, choice D, wrote a product, and 4 provided the correct product, $6 \times 24 = 144$. We believe that those students who did not write a product, or who wrote an incorrect product, may still reason multiplicatively. By contrast, students who calculated the correct product were providing evidence of their ability to recognize a multiplicative situation and apply multiplicative reasoning.

Summarizing Pedagogical Content Knowledge for Essential Understandings 1*a*, 1*c*, and 1*d*

Teaching the mathematical ideas in this chapter requires specialized knowledge related to the four components discussed in the Introduction: learners, curriculum, instructional strategies, and assessment. The four sections that follow summarize some examples of these specialized knowledge bases in relation to Essential Understandings 1*a*, 1*c*, and 1*d*. Although we separate them to highlight their importance, we also recognize that they are connected and support one another.

Knowledge of learners

When students are given the opportunity to make sense of problems, they typically approach problem solving in more than one way. Knowledge of learners involves anticipating likely and possible student responses. However, on some occasions, your students may demonstrate thinking that you have not anticipated or that you quickly recognize as unique. Collecting different student approaches to mathematics problems or common missteps can help you plan future instruction.

Figures 1.6–1.11, for instance, include diagrams created by students to represent 7×3, providing evidence of an inability to represent a multiplication situation, although the students could correctly supply the corresponding multiplication fact. For example, both Aaron (see fig. 1.10), who used roman numerals, and Jane (see fig. 1.11), who used shapes, created representations that were closely linked to the symbolic expression 7×3 instead of serving as pictorial representations of 7 groups of 3. Such representations may demonstrate a student's superficial understanding of the meaning of multiplication or inability to reason multiplicatively. Students' efforts to explain their reasoning or represent it pictorially or in a symbolic expression may also reveal that they are using additive reasoning in a multiplicative situation, as illustrated by Corinne's work, shown in figure 1.14.

Knowledge of curriculum

Many textbooks introduce multiplication as repeated addition, showing, for example, that $3 \times 5 = 5 + 5 + 5$, but such a narrow definition of multiplication limits the transfer of multiplicative reasoning and the development of deeper understanding of multiplication. For example, students who view multiplication as repeated addition may struggle to develop an appropriate understanding of the multiplication of fractions. Facing the need to compute $^2/_3 \times 5$, for example, they are likely to find it difficult to imagine what it means to add 5 two-thirds times. Continuing to focus on multiplication as repeated addition creates unnecessary confusion for students at later grade levels. How do the lessons in your curricular materials help students build their understanding of the meaning of multiplication, the multiplicative unit, and the multiplier? The task probing students' understanding of 7 groups of 3 and 7×3, shown in figure 1.6, provides opportunities for students to draw diagrams to demonstrate their understanding. How do your curricular materials emphasize the drawing and interpretation of diagrams?

Knowledge of instructional strategies

Teachers have a multitude of instructional strategies to draw on in teaching multiplication; this section highlights a few examples. One strategy is to ask students to interpret diagrams that focus on unmeasured multiplicative relationships. Figure 1.12 shows one such task, calling for a comparison of two rectangles of different lengths. Encouraging students to make comparisons is a good general strategy for helping students recognize and understand connections. Fosnot and Dolk (2002) suggest that students should wonder, ask questions (such as Why? and What if?), and notice patterns. Facilitating a whole-class discussion in which the children compare and contrast the problems in figure 1.1 could generate interesting ideas and questions from the children.

An instructional strategy that is useful in teaching multiplication as well as many other topics in mathematics involves having students analyze responses from fictitious students. Figure 1.13 shows the Star Sticker scenario, and students are asked to agree or disagree with conjectures about a multiplicative scenario from three fictitious children. This strategy can also lead to interesting whole-class discussions. In this case, one of the fictitious conjectures involves an additive comparison in a multiplicative context—a misunderstanding that teachers in grades 3–5 commonly encounter. Rather than introduce this misunderstanding through the work of a child in the classroom, you can highlight this idea and initiate a discussion of it through an examination of the work of "other children." Students can then justify their reasoning for their agreement or disagreement with work that some other student has supposedly done.

Knowledge of assessment

As noted in the Introduction, Wiliam (2007) emphasizes the importance of selecting tasks that provide opportunities for students to demonstrate their thinking and for teachers to make instructional decisions. The work that students generated in response to the tasks discussed in this chapter provides evidence about their understanding and misunderstanding of multiplication. The tasks ask students to explain their thinking, giving their teachers more evidence to interpret, assess, and use in making determinations about future instruction. The multiple-choice item displayed in figure 1.17 about a number sentence to use to find the number of Marlene's muffins asks students not only to choose the number sentence that represents the situation but also to explain their choice. The choice itself provides information about students' ability to give correct or incorrect responses. However, the requirement to explain the rationale for the choice provides teachers with additional information about what the students understand. Yet, in many cases these explanations are likely to be insufficient or vague. Therefore, another assessment strategy is called for as a follow-up, perhaps involving questions such as—

- "Why did you select an expression with the multiplication sign?"

- "Why is $6 \times \underline{\hspace{1cm}} = 24$ an incorrect answer?"

- "What information in the problem did you use to determine the correct number sentence?"

This strategy may allow you to collect additional assessment information, moving the children's understanding forward, or challenging an identified misconception.

Conclusion

Otto and colleagues (2011) emphasize the importance of multiplication, as quoted earlier: "Multiplication is a fundamental operation that is used to solve everyday problems" (p. 10). Yet, many students struggle to develop a deep understanding of multiplication and the underlying ideas related to it. It is critical to help students develop essential understandings related to the multiplicative unit and the idea of applying a multiplier to iterate, or make multiple copies of, that multiplicative unit. Therefore, it is helpful during instruction to identify and discuss the multiplicative unit and the multiplier in specific multiplicative situations explicitly.

Developing these essential understandings requires a careful selection of tasks and effective questions, yet this effort lays the foundation for developing essential understanding related to using multiplication skillfully in problem solving. Effective instruction about the use of multiplication in problem solving involves knowing about multiplication problem types, representations, and strategies, and these are the focus of the next chapter.

into practice

Chapter 2
Problem Types, Representations, and Strategies

Essential Understanding 1*b*
Each multiplicative expression developed in the context of a problem situation has an accompanying explanation, and different representations and ways of reasoning about a situation can lead to different expressions.

Applying multiplication appropriately in the context of a problem is an essential mathematical skill. Using multiplication effectively in problem solving depends on being able to create, understand, and explain an accompanying symbolic expression or other representation depicting the situation in the problem. Helping students build on their informal understanding, develop visual models for multiplicative thinking, and connect these models to the formal symbols for multiplication and division is critical to ensuring that they build a strong foundational understanding of the idea presented as Essential Understanding 1*b* in *Developing Essential Understanding of Multiplication and Division for Teaching Mathematics in Grades 3–5* (Otto et al. 2011).

This chapter begins by examining the various types of contextual problems (also called *word problems*) for multiplicative situations, including the problem types that division could be used to solve. The chapter then examines possibilities for using representations to build students' understanding of these contextual situations. Last, the chapter explores strategies that students use for solving problems in these contexts and briefly discusses ways of promoting the use of more sophisticated strategies for the basic facts for multiplication and division.

Working toward Essential Understanding 1*b*

Problems set in contextual situations provide a starting point for introducing and extending student understanding of multiplication and division. Contextual problems are a good place to begin because students are most familiar with multiplicative situations that they encounter outside classroom mathematics. Students have typically seen objects arranged and grouped in different ways, providing a natural connection to multiplicative reasoning.

In addition to providing students in grades 3–5 with contextual problems to solve, teachers can assess their understanding of multiplication by asking them to generate their own problems. For example, we asked a group of third-, fourth-, and fifth-grade students to write a word problem that would require computing 7 × 9 to determine the solution. Figure 2.1 shows one representative problem from a student in each grade level, 3–5. Reflect 2.1 guides you in comparing and contrasting these problems created by students.

Reflect 2.1

Identify the multiplicative unit (multiplicand) and the number of multiplicative units (multiplier) for the word problems generated by three students, shown in figure 2.1.

How are the problems similar to or different from one another?

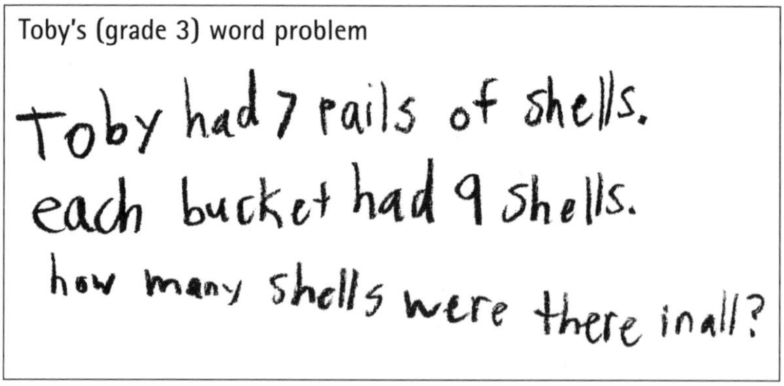

Toby's (grade 3) word problem

Toby had 7 pails of shells.
each bucket had 9 shells.
how many shells were there in all?

Fig. 2.1. Three examples of word problems requiring computing 7 × 9, written by students in grades 3–5

Bob's (grade 4) word problem

bob had 7 snakes in his house

billy had 9 times more snakes then bob many snakes does billy have?

Luke's (grade 5) word problem

Luke wants to give 7 treats in little bags. He wants to give these to 9 of his friends. How many treats does luke need to get.

Fig. 2.1. *Continued*

Both Toby and Luke wrote problems involving equal groups. Toby used a bucket containing 9 shells as the multiplicative unit and 7 buckets as the number of multiplicative units in his problem. Luke used a bag containing 7 treats as the multiplicative unit and 9 bags as the multiplier in his. Bob, however, created a different type of word problem. Bob compared the number of snakes he had with the number that his friend, Billy, had. In this case, one interpretation is that Bob's 7 snakes represent the multiplicative unit. In all, 9 iterations of this unit are necessary to determine the number of snakes that Billy has. On one hand, Bob's word problem might reflect an understanding of multiplicative comparison, but on the other hand, it might indicate a superficial knowledge demonstrated by applying a literal translation of "7 times 9."

Problem types for multiplication and division

Greer (1992) identified four different types of situations that involve multiplication and division:

1. Equal groups

2. Multiplicative comparison

3. Cartesian product

4. Rectangular area

Van de Walle (2007) noted that the first two types—equal groups and multiplicative comparison—provide a common entry point into multiplication and division in elementary school. He also emphasized various subtypes of multiplicative situations that involve equal groups and multiplicative comparison. Figure 2.2 identifies these subtypes and presents examples; Reflect 2.2 offers questions to frame your thinking about these problem types.

Reflect 2.2

Compare and contrast the problem types that are named and associated with examples in figure 2.2.

For each example, identify the multiplicative unit and the number of multiplicative units (multiplier).

Which problem types would you anticipate to be more or less difficult for your students?

Problem Types	Examples
Equal Groups	
Equal Groups: Result Unknown	Dusty has 4 bags of candy. Each bag has 6 pieces of candy. How many pieces of candy does Dusty have? Pat has 4 ribbons. Each ribbon is 6 inches long. How many inches of ribbon does Pat have?
Equal Groups: Size of Multiplicative Unit Unknown (Also known as *partitive* division situations)	Dusty has 24 pieces of candy. He wants to put the same number of pieces of candy into each of the 4 bags that he has and use all of the candy. How many pieces of candy should he put into each bag? Pat has 24 inches of ribbon. She wants to use all the ribbon and cut it so there are 4 pieces that are equal in length. How long, in inches, should each piece be?
Equal Groups: Number of Multiplicative Units Unknown (Also known as *measurement* division situations)	Dusty has 24 pieces of candy. He put all of the pieces into bags with 6 pieces of candy in each bag. How many bags did Dusty use? Pat has 24 inches of ribbon. She wants to cut the ribbon to make pieces that are 6 inches long. How many pieces of ribbon can Pat make?

Fig. 2.2. Problem types involving multiplicative reasoning (Greer 1992; Van de Walle 2007)

Problem Types	Examples
Comparison	
Comparison: Result Unknown	Marcia has 6 pieces of candy. Jacob has 4 times as many pieces of candy as Marcia. How many pieces of candy does Jacob have? Pat has 4 inches of ribbon. Michelle has 6 times the length of ribbon that Pat has. How long, in inches, is Michelle's ribbon?
Comparison: Size of Multiplicative Unit Unknown (Also known as *partitive* division situations)	Jacob has 24 pieces of candy. He has 4 times the number of pieces of candy that Marcia has. How many pieces of candy does Marcia have? Michelle has 24 inches of ribbon. She has 6 times the length of ribbon that Pat has. How long, in inches, is Pat's ribbon?
Comparison: Number of Multiplicative Units Unknown (Also known as *measurement* division situations)	Jacob has 24 pieces of candy. Marcia has 6 pieces of candy. How many times as many pieces of candy does Jacob have than Marcia? Michelle has 24 inches of ribbon. Pat has 4 inches of ribbon. Michelle has how many times the length of ribbon that Pat has?

Fig. 2.2. *Continued*

Problems of these types always involve three quantities: the multiplicative unit, the number of multiplicative units, and the product. These situations provide students with two of the three quantities and ask them to find the missing quantity. Students in grades 1 and 2 can solve some problems of these types by modeling the situations directly, as discussed later in this chapter. However, until students have experiences with multiplicative reasoning, working with comparison problems can be difficult. Students will have little understanding of the meaning of "5 times as much," for example. However, teachers can introduce these types of problems by using words like "doubling" or "tripling" and connecting these terms to "two times as much" and "three times as much."

Students are often more familiar with contexts involving equal groups, as is evident when they are asked to write word problems related to multiplication; recall the examples in figure 2.1. Of the 87 students who were asked to write word problems that could be solved by computing 7 × 9, 63 were able to generate such a problem. The majority of these students ($^{52}/_{63}$, or 83 percent) provided problems of the equal groups type. Fewer students offered multiplicative comparison problems ($^{11}/_{63}$, or 17 percent)—only about one-fifth as many as offered equal groups problems.

Students in this group did not use the other problem types of multiplicative word problems (that is, Cartesian product and rectangular area), probably because experiences with multiplication in grades 3–5 focus on problems of the equal groups and multiplicative comparison problem types.

Physical and symbolic models for multiplication and division

The effective use of physical models and representations is a key component of successful mathematics teaching (Clarke and Clarke 2004). Thus, developing in students a deep understanding of multiplication and division requires considering the various representations and interpretations of these models. However, using various models calls for caution, since research demonstrates the difficulties that surface when students do not develop a deep understanding of the mathematical ideas that underlie the models and representations that are not inherent in them (Ball 1992; Gravemeijer et al. 2002).

In your work in elementary classrooms, you have probably encountered students who demonstrate understanding of mathematical ideas related to multiplication and division in a particular context or when using a specific model but are unable to exhibit that understanding in other situations. To help students visualize multiplicative relationships for the problem types identified above, you must carefully consider the representations that you use in your classroom. Each representation provides different insight into the multiplicative structure and can clarify particular properties of multiplication and division (see Chapter 4, "Multiplication and Division Properties"). You have the critical responsibility of assessing your students' current knowledge related to the big ideas of multiplication and division as well as recognizing the power and limitations of various representations that you use in the classroom.

This section explores three fundamental types of physical or pictorial models that can help students develop their understanding of the meaning of multiplication and division: (a) linear models (for instance, number lines, ribbons, and line segments), (b) arrays, and (c) discrete sets of objects. It is important to recognize that students can model the different problem types identified in figure 2.2 by using any one of these models. To begin exploring models of these three types, consider the questions in Reflect 2.3.

Reflect 2.3

Below is a problem of the type identified in figure 2.2 as Equal Groups: Result Unknown:

> Dusty has 4 bags of candy. Each bag has 6 pieces of candy. How many pieces of candy does Dusty have?

Show how students could model this situation by using (a) a number line, (b) an array, and (c) sets of objects.

You might use a model that you think captures or exemplifies a particular mathematical idea, but your students may not view the model as you do. Thus, you must regularly assess your students' interpretations of the models in use to understand whether their view demonstrates proper mathematical understanding. It is important that you carefully select the models for the examples that you use in the classroom. You may see that a particular model applies to all situations that are similar in some way, but your students may not see the same general applicability to the example that you do.

For the situation in the problem in Reflect 2.3, the multiplicative unit is 6 pieces of candy. The number line model shown in figure 2.3 represents this situation. The multiplicative unit is a distance of length 6 on the number line. Sometimes teachers and students refer to a multiplicative unit identified as a distance on the number line as a "jump," and they represent the number of multiplicative units by the number of jumps of this length on the number line—in this case, 4.

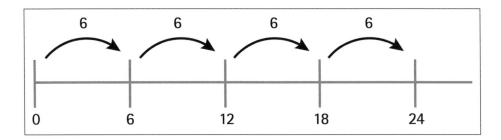

Fig. 2.3. A number line representing 4 jumps of 6

A second way to model the situation is to use an array. The standard method for using an array to model the situation of 4 bags of 6 pieces, or 4 × 6, is 4 rows with 6 objects in each row (see fig. 2.4). The number of rows represents the number of multiplicative units, and the number of objects in each row represents the size of the multiplicative unit. In figure 2.4, the loop around the first row emphasizes the multiplicative unit in the model. The entire array is 4 times larger than the top row.

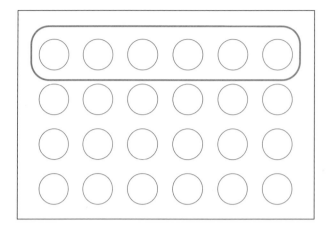

Fig. 2.4. An array with 4 rows and 6 objects in each row

A third way to model the situation of Dusty's bags of candy is to use sets of discrete objects. Modeling the situation in the problem in Reflect 2.3 involves creating 4 groups (number of multiplicative units) with 6 objects in each group (size of the multiplicative units). Figure 2.5 shows 4 ovals containing equal groups of 6 circles to represent the situation of 4 bags with 6 pieces of candy in each bag.

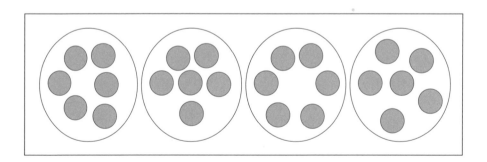

Fig. 2.5. An example of the set model for 4 groups of 6 objects

Students should be able to use multiple representations to model this situation, since each representation provides insight into the situation and reflects its meaning. However, some models may be easier than others for students to use for particular situations, and even in working with these, students may not attend to the multiplicative structure that the representation models for the situation.

For example, in the case of 4 bags of 6 pieces of candy, of the three representations shown in figures 2.3–2.5, the discrete representation in figure 2.5 models the situation most closely because seeing the groups is easiest in it. However, students may attend to the individual counts of objects within each group rather than the recognizing that the diagram shows 4 times as much as the original group of 6 objects. Some students may not identify the size of the group (the size of the multiplicative unit) in the row in the array model shown in figure 2.4 as easily as in one of the groups shown inside an oval in figure 2.5. It is easy to see the multiplicative unit and the number of multiplicative units in the number line model in figure 2.3. However, this can be a challenging model for students who have difficulty skip counting by sixes.

Although the set model may be best for this situation, both the array and number line models provide stronger connections with students' future work in multiplying fractions (see Chapter 5, "Algorithms for Multiplication and Division"). Students need to see the connections across these representations and recognize the multiplicative unit and the number of multiplicative units in each one. They also need to recognize that these models can represent a variety of multiplicative situations. For example, similar models can represent situations in problems of the types identified in figure 2.2 as Equal Groups: Result Unknown and Comparison: Result Unknown.

Another important aspect to consider is how to help your students connect these physical models with the symbolic representations for multiplication and division. As students begin to generate these representations, it is important that they encounter the formal mathematical symbols for multiplication and division and recognize their meaning. Being consistent in the symbolic expression that you use to refer to a model or situation will help students build their understanding. It is very important, for example, to write 4×5 consistently to stand for 4 groups of 5. Although 5×4 generates the same result as 4×5, many of the models for 4×5 differ from the models for 5×4. Suggesting to students that it does not matter whether they (or you) write 4×5 or 5×4 would be misleading, since they could then assume that $20 \div 4$ is the same as $4 \div 20$. Instead, clarifying that they are talking about 4 equal-sized groups of 5, written as 4×5, is important, ensuring that they can identify each group of 5 as well as the 4 groups in each physical representation that they create.

Suppose that the problem in Reflect 2.3 takes a slightly different form:

> Dusty has 24 pieces of candy. He wants to put the same number of pieces of candy into each of the 4 bags that he has and use all of the candy. How many pieces of candy should he put into each bag?

In this case, it becomes a problem of the type identified in figure 2.2 as Equal Groups: Size of Multiplicative Unit Unknown. Students should recognize that they can write the number sentence 4 × ? = 24, or 24 ÷ 4 = ?, since they know the number of groups and the total amount. Too often teachers ask, "What operation do we use in this situation?" when in fact students can use a variety of operations to solve the problem. Allowing students to share their strategies and the operations that they used, and helping them make connections among the operations, can assist them in developing their understanding of the meaning of the various operations and eliminate the misnomer "multiplication problem" for a problem set in certain types of situations.

Similarly, the problem in Reflect 2.3 can be recast as problem of the type identified in figure 2.2 as Equal Groups: Number of Multiplicative Units Unknown:

> Dusty has 24 pieces of candy. He puts all of the pieces into bags with 6 pieces of candy in each bag. How many bags does Dusty use?

Students should recognize that they can connect this problem with the symbolic representation ? × 6 = 24 or the expression 24 ÷ 6 = ? because they know the total amount and the size of the multiplicative units (or groups), but the number of multiplicative units (or groups) is unknown. Again, it is important to use notation consistently and emphasize the meaning of the notation for multiplication and division to help students develop a deep understanding of the mathematical symbols.

Textbooks often represent multiplicative situations with groups of objects, rectangular arrays, and number lines. Therefore, you should ensure that your students not only develop understanding of the meaning of these different models, but also make connections among them. However, problems in textbooks may be presented in a way that anticipates only one correct response. For example, a problem such as that in figure 2.6 may be found in textbooks for third-, fourth-, and fifth-grade students.

Fig. 2.6. A sample multiplication problem from a textbook

The second row of missing values ($\square + \square + \square = \square$) suggests that students are expected to interpret this situation as 3 groups, with 8 objects as the multiplicative unit. If not for this second row, students could also view this representation as 6 groups of 4, 12 groups of 2, 24 groups of 1, and 1 group of 24 (by far the easiest!). When using textbook tasks with students, be aware that some representations and instructions may be too vague, and others may be too restrictive. Furthermore, tasks such as this one can promote the view of multiplication as repeated addition rather than assist students in recognizing the multiplicative unit and the number of multiplicative units.

Student strategies for multiplication and division

Students encounter different problem types and models as they develop essential understandings related to multiplication and division. In addition, students in grades 3–5 use different strategies as they solve problems. Carpenter, Fennema, and Franke (1996) explain that students may directly model a situation by using counters or other manipulatives.

A next step may involve modeling the situation by using a pictorial representation, as illustrated by the sample of student work shown in figure 2.7 in response to the Bags of Oranges problem:

> Maria has 5 bags of oranges. Each bag has 7 oranges. How many oranges does she have?

In representing this problem pictorially, a third-grade student, Elizabeth, drew 5 larger circles to represent the bags of oranges and then drew 7 smaller circles inside the 5 larger circles to represent the individual oranges. She then counted each individual orange to determine the total of 35.

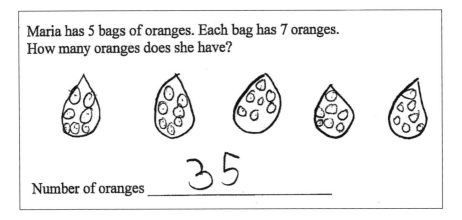

Fig. 2.7. Elizabeth's (grade 3) representation for the Bags of Oranges problem

In response to the same task, Jose, also a third grader, drew a diagram similar to Elizabeth's; figure 2.8 shows Jose's work. However, rather than count each orange to find the total number of oranges, he used a counting strategy—in this case, skip counting by sevens, as indicated by the numbers 7, 14, 21, 28, and 35 in his work.

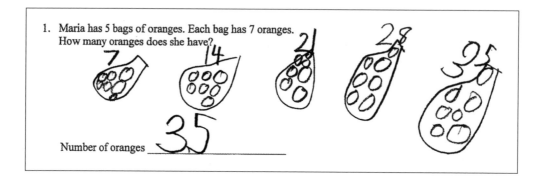

Fig. 2.8. Jose's (grade 3) representation for the Bags of Oranges problem

Brian, another third grader, also used skip counting to solve the problem. However, his work, pictured in figure 2.9, shows that he skip-counted by fives. Does Brian understand that the idea of the commutative property applies in this situation and that skip counting by fives is more efficient than skip counting by sevens?

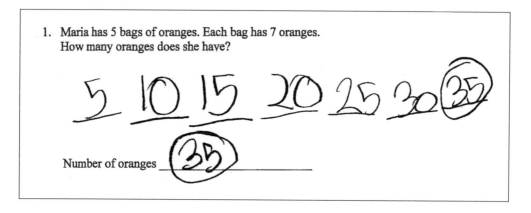

Fig. 2.9. Brian's (grade 3) skip-counting strategy for the Bags of Oranges problem

Brian's work does not make his thinking clear. Therefore, asking him to explain it or posing some additional tasks to determine why he skip-counted by fives could be helpful in assessing his understanding. For example, Brian could be asked to solve the following problem:

Kevin has 7 bags of oranges. Each bag has 5 oranges. How many oranges does he have?

Alternatively, having Brian draw a representation for the original problem about Maria's bags of oranges could be helpful. Brian could be asked where the 7 bags are, where the 5 oranges are in each bag, and where the total number of oranges is. Such questions would provide insight into Brian's views of the multiplicative unit, the number of multiplicative units, and the product.

The Common Core State Standards for Mathematics (CCSSM; National Governors Association Center for Best Practices and Council of Chief State School Officers [NGA Center and CCSSO] 2010) expects students in grade 3 to relate the concept of area to the operations of multiplication and addition. Figure 2.10 provides the complete statement of this grade 3 expectation in CCSSM.

Common Core State Standards for Mathematics, Grade 3

Use place value understanding and properties of operations to perform multi-digit arithmetic.

7. Relate area to the operations of multiplication and addition.

    a. Find the area of a rectangle with whole-number side lengths by tiling it, and show that the area is the same as would be found by multiplying the side lengths.

    b. Multiply side lengths to find areas of rectangles with whole number side lengths in the context of solving real world and mathematical problems, and represent whole-number products as rectangular areas in mathematical reasoning.

Fig. 2.10. Relating area to multiplication. Measurement and Data, CCSSM 3.MD 7a, 7b (NGA Center and CCSSO 2010, p. 25)

To help students connect the idea of area with multiplication and to determine their understanding of the relationship, we gave third-grade students the Tile task, a problem involving a different type of multiplicative situation (shown in fig. 2.11). This task shows the students an array that they may have difficulty viewing in a multiplicative manner, since part of it is covered. Examine the Tile task in figure 2.11 and consider the question in Reflect 2.4 about what students' responses might show about their understanding of the relationship between area and multiplication.

Reflect 2.4

How might a student demonstrate multiplicative reasoning when solving the Tile task, displayed in figure 2.11?

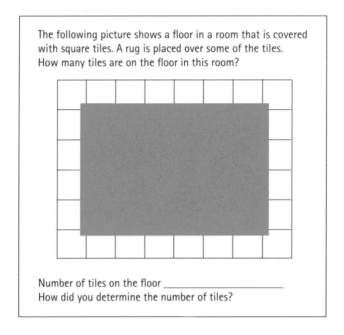

The following picture shows a floor in a room that is covered with square tiles. A rug is placed over some of the tiles. How many tiles are on the floor in this room?

Number of tiles on the floor _____

How did you determine the number of tiles?

Fig. 2.11. The Tile task

Figures 2.12–2.15 exhibit the work of four third-grade students who completed the Tile task. Consider the questions in Reflect 2.5 as you examine the samples of student work in the figures.

Reflect 2.5

Which students appear to use multiplicative reasoning?

How could you help students identify a multiplicative unit in this task?

What questions would you ask these students to help them move forward?

Number of tiles on floor ___57___

How did you determine the number of tiles? I made a picture to help the I counted and I got 57 as my number of tiles on the floor.

Fig. 2.12. Lydia's (grade 3) work for the Tile task

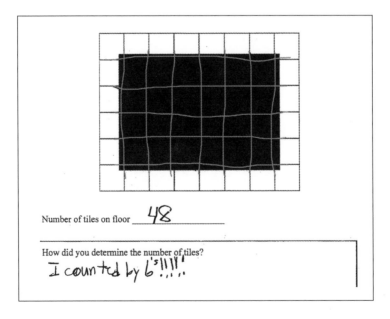

Number of tiles on floor ___48___

How did you determine the number of tiles?
I counted by 6's!!!!!..

Fig. 2.13. Mark's (grade 3) work for the Tile task

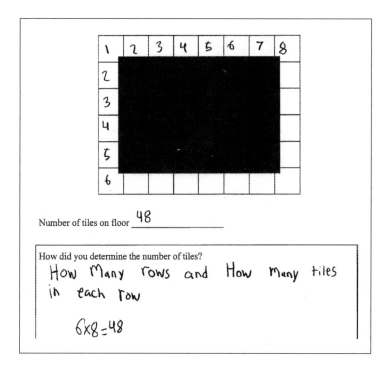

Number of tiles on floor ___48___

How did you determine the number of tiles?

How Many rows and How Many tiles in each row

6×8=48

Fig. 2.14. Chase's (grade 3) work for the Tile task

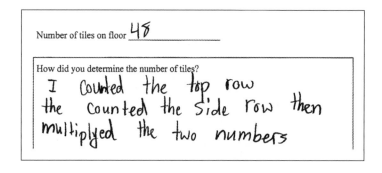

Number of tiles on floor ___48___

How did you determine the number of tiles?

I Counted the top row the Counted the side row then multiplyed the two numbers

Fig. 2.15. Shelby's (grade 3) work for the Tile task

The majority of the 54 third graders who completed the Tile task counted the tiles one by one, often drawing lines to show those tiles under the rug. Students who counted the tiles were not using multiplicative reasoning, because they did not identify a multiplicative unit or consider the iterations of the multiplicative unit. Some students like Lydia, whose work appears in figure 2.12, did not draw the lines correctly, or they miscounted. Mark, whose work appears in figure 2.13, also drew lines over the shading but used a more sophisticated counting strategy—skip counting—revealing multiplicative reasoning in identifying the multiplicative unit as a column of 6 tiles.

Both Chase, whose work appears in figure 2.14, and Shelby, whose work appears in figure 2.15, found the total number of tiles by using multiplication—a rarely chosen strategy among this group of third-grade students. Chase's explanation, referring explicitly to his goal of determining "how many rows and how many tiles in each row," indicates a high likelihood that he was using multiplicative reasoning, conceptualizing the number of tiles in each row as the multiplicative unit and the number of rows as the multiplier.

How did Shelby's reasoning differ from Chase's? First of all, whether she identified a multiplicative unit is not clear. Instead, Shelby might have learned that the numbers need to be multiplied in this type of situation. Although she appears to have known that multiplication would give her the answer, she might not know why. Further probing would be necessary to understand why she multiplied and what she saw as the multiplicative unit and number of multiplicative units.

In addition to direct modeling and counting strategies, Carpenter, Fennema, and Franke (1996) also discuss a student strategy that they refer to as *derived facts*. Derived facts for multiplication are based on what students know about addition, subtraction, and multiplication. They often rely on properties of these operations. For example, the sample of work in figure 2.16 illustrates how a third-grade student, Cheri, used a known fact, $7 \times 10 = 70$, to determine an unknown fact, 7×9. Cheri began with the known fact, $7 \times 10 = 70$, subtracted 7, and obtained $7 \times 9 = 63$. Her work also demonstrates a degree of awareness of the distributive property, which ensures that $7 \times 9 = 7 \times (10 - 1) = 7 \times 10 - 7 \times 1$.

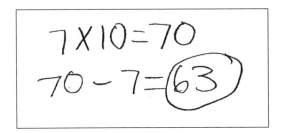

Fig. 2.16. Cheri's (grade 3) use of the derived facts strategy to compute 7×9

The third-grade student whose work is shown in figure 2.17 shifted from skip counting to derived facts, establishing 7 × 9 = 63 by using doubles (9 + 9 = 18, 18 + 18 = 36) to calculate the product for 7 groups of 9. Using derived facts is more efficient than either counting or direct modeling. Some students become so proficient in using derived facts that it is difficult to distinguish a derived answer from an answer recalled from memory.

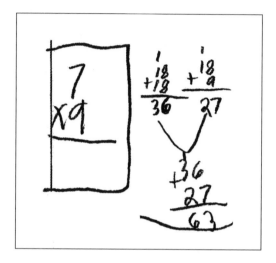

Fig. 2.17. Derived facts strategies for 7 × 9 (grade 3)

Summarizing Pedagogical Content Knowledge to Support Essential Understanding 1*b*

Teaching the mathematical ideas in this chapter requires specialized knowledge related to the four components presented in the Introduction: learners, curriculum, instructional strategies, and assessment. The four sections that follow summarize some examples of these specialized knowledge bases in relation to Essential Understanding 1*b*. Although we separate them to highlight their importance, we also recognize that they are connected and support one another.

Knowledge of learners

As illustrated by the samples of work on the Bags of Oranges task (see figs. 2.7–2.9) and the Tile task (see figs. 2.12–2.15), students may use various strategies when faced with multiplicative situations. Initially, students approach these situations through *direct modeling,* using physical objects or their fingers to represent quantities and physically pushing them together (Carpenter, Fennema, and Franke 1996).

Or they may represent the situation through the use of a picture (see, for example, Elizabeth's picture of the Bags of Oranges problem in fig. 2.7).

Students may use *counting strategies*, such as counting all or skip counting—the process of counting by multiples of a number. The sample of Jose's work on the Bags of Oranges problem, shown in figure 2.8, associates drawings of bags of oranges with numbers in a skip-count sequence, revealing a transition from working with a picture (drawing all the bags and oranges) to skip counting (writing the cumulative count above each bag). This transition will be complete when the student skip-counts without using physical objects or pictures.

Knowledge of curriculum

It is likely that student experiences with multiplication and division in grades 3–5 will be limited to specific problem types (for example, equal groups, multiplicative comparison, and rectangular area) or subtypes (for instance, result unknown). The examples of problem types in figure 2.2 can help you classify the types that are in your curricular materials. Which problem types are emphasized in the materials that you use? How do the lessons in these materials help students build their understanding of the meaning of these different problem types? Do students have opportunities to create their own representations or discuss the meaning of the different components of a representation for these problem types? Selecting, enhancing, or designing appropriate mathematical tasks involves taking into account the contexts, units, and language in the tasks. Further, it requires careful consideration not only of students' understandings and misconceptions, but also of the mathematical models and representations that the tasks use and how students perceive them.

Knowledge of instructional strategies

Teachers can draw on many and varied instructional strategies in helping their students interpret multiplicative situations and work with models and representations of them. We highlight one example here. As you facilitate discussions about your students' representations, emphasize the meaning that lies behind their various components. As your students engage in a task, pose questions such as the following:

- "What does this [*using a gesture to indicate a specific component*] represent?"

- "Where is the multiplicative unit in your representation?"

- "How does your representation differ from your partner's? How is your representation similar to your partner's?"

Knowledge of assessment

Carefully designed assessment tasks will generate different student strategies, giving teachers information that they can use to help students develop accurate understandings and more sophisticated approaches. For example, figure 2.1 shows word problems that students themselves generated. By inspecting your students' problems, you will be able to determine which students selected an equal groups situation or comparative situation.

When students solved the Tile task, their teachers could assess which students were counting all and which students were recognizing the multiplicative unit. The design of the task, with some of the floor tiles covered, made these distinctions apparent. If all the tiles had been displayed, students would have been more likely to count them, and the resulting work would not have indicated how they had counted.

Conclusion

Helping your students develop a deep understanding of multiplication and division requires giving them opportunities to consider and work with problems of various types and with various models—number lines, arrays, and sets—as well as both pictorial and symbolic representations. Moreover, it is important to recognize and track which strategies students select and how their use of these grows in sophistication over time. As a result, you must carefully select tasks and pose effective questions to help your students develop essential understandings related to using multiplication in problem solving and lay the foundation for developing essential understanding of the meaning of division, which is the subject of the next chapter.

into
practice

Chapter 3
The Meaning of Division

Essential Understanding 1e
Division is defined by its inverse relationship with multiplication.

Essential Understanding 1f
Using proper terminology and understanding the division algorithm provide the basis for understanding how numbers such as the quotient and the remainder are used in a division situation.

The concepts of division and multiplication are closely related, as highlighted in *Developing Essential Understanding of Multiplication and Division for Teaching Mathematics in Grades 3–5* (Otto et al. 2011). Students who have an understanding of the meanings of division, the associated terminology, and the relationship between division and multiplication are prepared to solve problems and understand multiplicative situations. Communicating these ideas in ways that make them meaningful to students in grades 3–5 and assessing their understanding are the subjects of this chapter.

Working toward Essential Understandings 1e and 1f

Mulligan and Mitchelmore (1997) explain that the term *multiplicative* describes situations that lead to either multiplication or division and that every multiplication situation can lead to various division problems. Therefore, students can develop an understanding of the meaning of multiplication and division at the same time. Students can make connections between multiplication and division as they "think multiplication" when working with division. Students should not simply identify some problems as "multiplication problems" and others as "division problems." However, this view is at odds with the typical approach to multiplication and division in U.S. textbooks, which commonly introduce multiplication and division in separate chapters. When students focus on a single problem type or operation, they often limit their thinking to the "strategy of the day" while failing to make important connections with other ideas.

We observed such difficulties when we provided a task that directed third-grade students to construct 7 groups of 3 circles. Because the students had recently modeled sharing situations, approximately 15 percent of them automatically approached the task as a sharing problem, providing a response similar to that shown in figure 3.1.

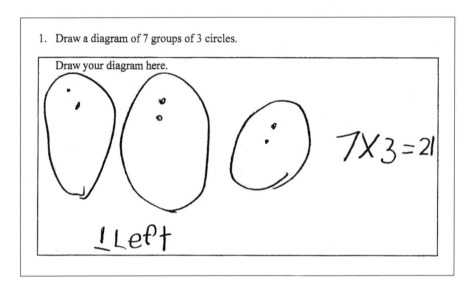

Fig. 3.1. A third-grade student applying a "strategy of the day" in an inappropriate context

Although the student whose work is shown in the figure modeled the situation incorrectly by attempting to distribute 7 objects in 3 equal groups, she wrote a correct number sentence for the situation: $7 \times 3 = 21$. Such disconnects between pictorial and symbolic representations occur frequently when problems of a single type are the primary focus of separate lessons. As Rohrer and Pashler (2010) note, varying problem structures within lessons can increase long-term retention and understanding. Diversifying the problems that you give your students compels them to focus on the deeper structure of the multiplicative situations that the problems present rather than look for key words or other superficial aspects.

While emphasizing the need to vary the types of multiplicative (and additive) situations that students encounter in problems, this chapter also focuses on helping students build their understanding of the meaning of division. A useful starting point is the structure of contextual problems that can be modeled by the operation of division. Reflect 3.1 launches this discussion with a guided examination of the two word problems in figure 3.2.

Reflect 3.1

Draw a representation that models Marcia's Bananas and Tabitha's Bananas, the two problems in figure 3.2.

Compare and contrast the problems. How are they similar to or different from each other?

What type of reasoning would students be likely to use in approaching each of these situations?

Marcia's Bananas
Marcia has 40 bananas. She wants to put the bananas into bundles with 5 bananas in each bundle. How many bundles can she make?

Tabitha's Bananas
Tabitha has 40 bananas. She wants to put the bananas into 5 bundles with the same number of bananas in each bundle. How many bananas should Tabitha put in each bundle?

Fig. 3.2. Two contextualized problems that can be modeled by division

The first problem, Marcia's Bananas, gives the total amount (40 bananas) and the size of the multiplicative unit (5 bananas in a bundle) and asks for the number of multiplicative units (number of bundles). The second problem, Tabitha's Bananas, gives the total amount (40 bananas) and the number of multiplicative units (5 bundles) and asks for the size of the multiplicative unit (the number of bananas in a bundle).

Typically, problem situations that give the total number and the number of multiplicative units (or number of groups) and seek the size of the multiplicative unit (or the number in each group), like Tabitha's Bananas, are called *partitive division* situations. Such situations are also known as *partition division*, or *fair-sharing division* situations.

By contrast, situations that give the total amount and the size of the multiplicative unit (that is, the number in each group) and seek the number of multiplicative units, like Marcia's Bananas, are called *measurement division* situations. Such situations are also referred to as *repeated subtraction* or *quotitive division* situations.

Partitive division and measurement division have subtle differences that adults recognize as situations in which they can either use division or "think multiplication." However, students may not notice or understand these subtleties. As a result, they may apply different strategies initially when modeling them. Figures 3.3 and 3.4 present models that students created to represent 20 ÷ 4. Consider the question in Reflect 3.2 as you examine these two students' work.

Reflect 3.2

Figures 3.3 and 3.4 show two students' representations of 20 ÷ 4.

Which representation do you think would be more challenging for students to make?

How could the students' work on 20 ÷ 4 shown in the figures translate to their work with other problems and models?

Fig. 3.3. A number line representation of 20 ÷ 4

Fig. 3.4. An equal groups representation of 20 ÷ 4

The student whose work appears in figure 3.3 applied a measurement meaning of division by making equal jumps of length 4 backward and determining the number of jumps. This student interpreted 4 as a multiplicative unit. Similarly, when working with the situation in Marcia's Bananas in figure 3.2, students might start with the 40 bananas that Marcia has and repeatedly "subtract out" the multiplicative unit that Marcia has decided on (5 bananas to a bundle) until they had accounted for all of Marcia's bananas. Then they would need to count the number of multiplicative units that they had subtracted. A problem that has this structure, with the number of multiplicative units as the unknown quantity, is generally fairly easy for students to model by using objects or diagrams of objects.

The student whose work is shown in figure 3.4 applied a partitive meaning of division in working with $20 \div 4$, creating 4 circles to represent the multiplicative units and partitioning 20 tally marks equally among them. Faced with situations such as that in Tabitha's Bananas, shown in figure 3.2, students typically use a number of different strategies. For example, some students would try to model this situation by using a trial-and-error approach. Knowing that Tabitha has a total of 40 bananas and wants to make 5 bundles, they guess a number of bananas as a multiplicative unit and check to see whether creating equal units of that size exhausts the total number of bananas. If not, the students make adjustments until they create equal groups.

Other students might use a more sophisticated method to approach Tabitha's Bananas, modeling the situation by applying a strategy much like dealing cards, to give the same number of bananas to each multiplicative unit. For example, the students might draw 5 circles to represent the 5 bundles of bananas that Tabitha wants to make. Then they might draw one dot in each of the circles in turn to represent one banana. They would continue to "deal" dots until they reached the total count of 40. Other students might determine that dealing one dot at a time was not efficient and deal a larger number— for example, 2 or 4 dots. A problem of the partitive type, like Tabitha's Bananas, in which the size of the multiplicative unit is the unknown quantity, is generally more difficult for students than a problem of the measurement type, like Marcia's Bananas, in which the number of multiplicative units is the unknown quantity. Reflect 3.3 offers an opportunity to explore the structures of problems of these different types by creating examples.

Reflect 3.3

Write a word problem that students could solve by computing $20 \div 4$ while using *measurement division*.

Write another word problem that students could solve by computing $20 \div 4$ while using *partitive division*.

It is critical that students build an understanding of the meaning of these two different problem types—measurement division and partitive division. The Common Core State Standards for Mathematics (CCSSM; National Governors Association Center for Best Practices and Council of Chief State School Officers [NGA Center and CCSSO] 2010) recommend that third-grade students interpret division situations in these two different ways; CCSSM's statement of the standard appears in figure 3.5. Using the example of 56 ÷ 8, CCSSM asserts that students should be able to interpret the quotient as representing either the *number of equal-sized groups* into which 56 has been fair-shared or measured out or the *size of each equal-sized group* formed by partitioning 56 into 8 such groups.

Common Core State Standards for Mathematics, Grade 3

Interpret whole-number quotients of whole numbers, e.g., interpret 56 ÷ 8 as the number of objects in each share when 56 objects are partitioned equally into 8 shares, or as a number of shares when 56 objects are partitioned into equal shares of 8 objects each. *For example, describe a context in which a number of shares or a number of groups can be expressed as 56 ÷ 8.*

Fig. 3.5. Interpreting quotients in partitive and measurement ways. Operations and Algebraic Thinking, CCSSM 3.OA.2 (NGA Center and CCSSO 2010, p. 23)

We asked fourth- and fifth-grade students to write a word problem that would be solved by computing 20 ÷ 4. Figures 3.6–3.9 show problems that four fifth-grade students wrote. Examine the students' work, using the questions in Reflect 3.4 to guide your exploration.

Reflect 3.4

Examine the four problems written by fifth-grade students, shown in figures 3.6–3.9.

Which students appear to understand the meaning of division?

Which problem type (measurement or partitive) did each student create?

Write your word problem here.

Kadin has 20 Pennies. One day he decided to sort them into groups of 4. How many groups does he have. (A= 5 groups)

Fig. 3.6. Kadin's (grade 5) word problem for 20 ÷ 4

Write your word problem here.

Sara had 20 crayons, Sara used 4 crayons. How many crayons did sara did sara have left

crayons not used=5crayons

Fig. 3.7. Sara's (grade 5) word problem for 20 ÷ 4

There was 20 cookies. There we 4 kids each kid wanted the same amount of cookies. How many cookies did each kid get?

Fig. 3.8. Perry's (grade 5) word problem for 20 ÷ 4

Alina has 20 pencils. She has 4 friends and they all need some. How many pencils will each friend get?

Fig. 3.9. Alina's (grade 5) word problem for 20 ÷ 4

Half of the 124 students in our groups of fourth and fifth graders did not provide any word problem, and 4 students wrote a word problem involving subtraction. One of those students was Sara, the fifth grader whose work is shown in figure 3.7. Sara appears to know that 20 ÷ 4 = 5, but she does not present a situation that involves division.

Of the remaining students, 24 wrote word problems involving measurement division, similar to Kadin's problem, shown in figure 3.6. Thirty-four students wrote word problems involving partitive division, including Perry and Alina, whose work appears in figures 3.8 and 3.9, respectively. Perry's inclusion of the condition that "each kid wanted the same amount of cookies" indicates that he recognizes that the multiplicative unit—the number of cookies for each child—is constant. This is an important aspect of multiplicative reasoning. Alina, on the other hand, specifies no such condition. One could argue that she intended that each friend would get the same number of pencils, but that is not clear. As Alina's problem is written, Alina's friends could all get different numbers of pencils (such as 10, 5, 3, and 2), and the unequal distribution could still satisfy the conditions in the problem.

Problems related to division should also include situations that involve remainders. Students need to recognize that remainders may be handled differently in different types of situations. CCSSM expects fourth-grade students to be able to "solve multistep word problems posed with whole numbers and having whole-number answers using the four operations, including problems in which remainders must be interpreted" (4.OA.3, p. 29). Figure 3.10 presents three tasks involving remainders. Examine the tasks, using the questions in Reflect 3.5 to guide your investigation.

Reflect 3.5

Examine the Balloons, Birds, and Cookies tasks in figure 3.10.

What ideas about remainders are these tasks designed to assess?

What do students need to understand about remainders in multiplicative situations?

A group of 143 students from grades 3–5 completed the tasks shown in figure 3.10. Figure 3.11 contains the work of three students at different grade levels on the Balloons task. Examine their work before responding to the questions in Reflect 3.6.

Balloons

A mother had 20 balloons. She wanted to give them to her 3 children so that each child would have the same number of balloons. How many balloons did each child get?

Birds

A pet store owner has 14 birds and some cages. She will put 3 birds in each cage. How many cages will she need to use?

Cookies

A father has 17 cookies. He wants to give them to his 3 children so that each child has the same number of cookies. How many cookies will each child get?

Fig. 3.10. The Balloons, Birds, and Cookies tasks given to 143 third-, fourth-, and fifth-grade students

Reflect 3.6

How did the three students whose work appears in figure 3.11 deal with the remainder in the Balloons task, shown in figure 3.10?

Which students appear to have a correct understanding of remainders related to this situation?

How would you assist those students who did not demonstrate a correct understanding of remainders?

The Balloons task gives the total number of objects—20 balloons—and the number of multiplicative units—3 groups of balloons (for the 3 children)—and asks for the size of the multiplicative unit—the number of balloons in a group for each child. Because the total number of objects—20 balloons—cannot be partitioned into 3 equal-sized groups with no balloons left over, students must make the largest equal-sized groups that they can and decide what to do with the leftover balloons.

Both Gordon (grade 3) and Eugene (grade 4) drew diagrams to solve the Balloons problem. Rachel (grade 5) offered symbolic statements using both division and multiplication, making it clear that 2 balloons remain after each child receives 6, and she explicitly suggested that the mother could keep the remainder. Gordon, by contrast, did not give any explicit indication of the existence of a remainder, although he did indicate that each child would get 6 balloons. When a total number of discrete objects, such as balloons, cannot be subdivided, discarding the remainder may be appropriate. Eugene decided that two children would get 7 balloons, a choice that is contrary to the mother's wish that each child receive the same number of balloons. Thus, at least in this situation, Eugene does not appear to recognize the importance of the fixed size of the multiplicative unit—the equal-sized groups—whereas Rachel and Gordon do.

Fig. 3.11. Three students' work on the Balloons task

Figure 3.12 shows the work of three students—two fourth graders and one fifth grader—on the Birds task in figure 3.10. Examine these students' work, using the questions in Reflect 3.7 to guide your thinking.

Reflect 3.7

How did the three students whose work appears in figure 3.12 deal with the remainder in the Birds task, shown in figure 3.10?

Which students appear to have a correct understanding of remainders related to this situation?

How would you assist those students who did not demonstrate a correct understanding of remainders?

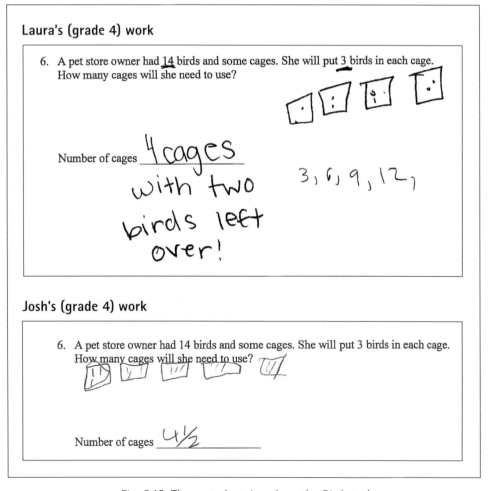

Fig. 3.12. Three students' work on the Birds task

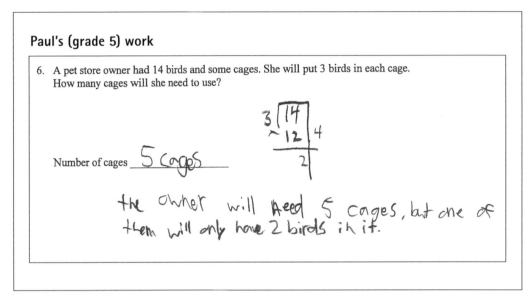

Paul's (grade 5) work

6. A pet store owner had 14 birds and some cages. She will put 3 birds in each cage. How many cages will she need to use?

Number of cages ___5 cages___

the owner will need 5 cages, but one of them will only have 2 birds in it.

Fig. 3.12. *Continued*

Laura (grade 4), Josh (grade 4), and Paul (grade 5) all agreed that the pet store owner will need at least 4 cages. Although Laura and Josh used diagrams to show their reasoning, Paul used a procedure to find the quotient and the remainder. Laura acknowledged a remainder of 2 birds but in effect discarded it. Although discarding the remainder is an appropriate strategy in the Balloons situation, it is inappropriate in the Birds context. Paul, by contrast, pointed out that the owner will need 5 cages, but one cage will only contain 2 birds. Josh's answer of $4\frac{1}{2}$ cages introduces a partial cage, an idea that is problematic in the context of the problem. The pet store owner will almost certainly need to have all the birds in cages, and a partial cage would seemingly be useless or meaningless in the situation, in which it would be more natural to "round up" as Paul did.

Figure 3.13 presents the work of three third-grade students on the Cookies task in figure 3.10. Examine these students' work while considering the questions in Reflect 3.8.

Reflect 3.8

How did the three students whose work appears in figure 3.13 deal with the remainder in the Cookies task, shown in figure 3.10?

Reflect 3.8. *Continued*

Which students appear to have a correct understanding of remainders related to this situation?

How would you assist those students who did not demonstrate a correct understanding of remainders?

Elijah's (grade 3) work

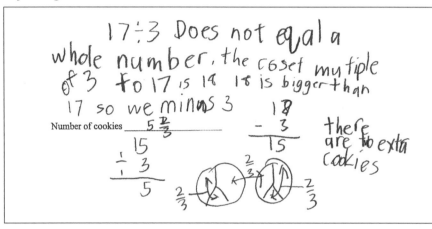

Allie's (grade 3) work

Fig. 3.13. Three students' work on the Cookies task

Rintaro's (grade 3) work

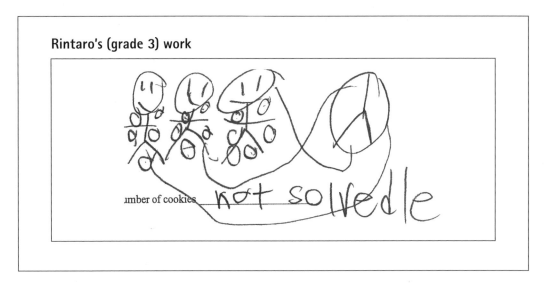

imber of cookies ____ not solvedle

Fig. 3.13. *Continued*

All three third graders, Elijah, Allie, and Rintaro, agreed that each student will get at least 5 cookies. However, Rintaro declared that this problem is not solvable, probably because he was unsure of how to deal with the remaining cookies. Allie decided to give them to the father in the same way that Rachel suggested that the mother take the remaining balloons in the Balloons task. In the case of cookies, however, a single cookie can sometimes be subdivided. Elijah illustrated this possibility in his representation by partitioning each remaining cookie into three parts and giving two parts to each child. Provided that the cookies can be partitioned in this manner, doing so is probably the most appropriate way to deal with the remainder in the Cookies context.

Taken as a group, the Balloons, Birds, and Cookies tasks highlight the fact that remainders are handled in different ways, not all of which are appropriate in every situation. Instead, students need to consider the context of the problem and choose a sensible method.

To help your students develop an understanding of the various ways to deal with remainders, have them examine various contexts, such as those presented here, and discuss their results with one another. Problems of the measurement type, like the Birds task, often lend themselves to "rounding up." Partitive problems, by contrast, may require that the remainder be discarded or that a partial quantity—like $2/3$ of a cookie—be considered.

Furthermore, by having your students work with remainders, you can help them make a key connection between division and fractions. In the Cookies context, the remaining 2 cookies could perhaps be shared fairly by the 3 children, giving each child $2/3$ of a cookie in addition to 5 whole cookies, with the key connection between division and fractions residing in the fact that $2 \div 3 = 2/3$. Grasping this specific fact can promote students' understanding of the more general fact that the quotient $a \div b$ is equivalent to the fraction a/b.

Summarizing Pedagogical Content Knowledge for Essential Understandings 1e and 1f

Teaching the mathematical ideas in this chapter requires specialized knowledge related to the four components introduced in the Introduction: learners, curriculum, instructional strategies, and assessment. The four sections that follow summarize some examples of these specialized knowledge bases in relation to Essential Understandings 1e and 1f. Although we separate them to highlight their importance, we also recognize that they are connected and support one another.

Knowledge of learners

Some students have misconceptions related to the meaning of remainders. They may learn to write, for example, "4 R2" in earlier grades or "$4^1/_2$" in later grades, yet in problem situations involving discrete objects such as birdcages, this handling of remainders is inappropriate. This is an important misconception to anticipate and challenge as you help your students build their understanding of the meaning of remainders in problem situations.

Knowledge of curriculum

An analysis of your curricular materials can help you determine which problems to use and how to sequence them. As this chapter has demonstrated, it is important to vary problem types so that students can see connections between multiplication and division problems. Students should also have opportunities to compare and contrast tasks such as the measurement and partitive Bananas problems in figure 3.2 and the Balloons, Birds, and Cookies problems involving remainders in figure 3.10. It is important to include tasks with different features, like partitive and measurement structures and different contexts for remainders.

Including curricular tasks that show disconnects in students' thinking is important. Faced with the task of drawing a diagram of 7 groups of 3 circles, the third grader whose work is shown in figure 3.1 applied a sharing strategy–the "strategy of the day." Although she attempted to "fair-share" 7 dots among 3 circles, she demonstrated that she is able to write the correct symbolic statement: $7 \times 3 = 21$. Given the task of writing a word problem that would be solved by computing $20 \div 4$, Sara, the fifth grader whose work is shown in figure 3.7, produced a subtraction problem.

Some students have limited experiences with modeling multiplicative situations and making sense of them through the use of diagrams. Including curricular tasks that require students to model different problem types such as those shown in figure 3.2 can build their understanding.

How are multiplication and division introduced in your curricular materials? Do your materials have a curricular sequence that makes sense and provides sufficient time for students to develop essential understanding? How do your curricular materials handle different division problem types and the use of remainders? How do your curricular materials make connections between pictorial and symbolic representations?

Knowledge of instructional strategies

Teachers have a multitude of instructional strategies to draw on in teaching multiplication and division. We highlight one example here. As students begin to work with multiplication and division situations, they need to use physical materials or draw diagrams to model these problems. For example, for the Cookies task, some students may need to model the situation directly with objects—perhaps 3 large circles cut from construction paper to represent plates for the 3 children, and 17 small construction-paper circles (or other small easily divisible objects) to represent cookies (Carpenter, Fennema, and Franke 1996). If your students directly model this situation or a similar one, you will be able to observe which students use a "dealing cards" strategy, doling out one cookie to a plate until the cookies are gone, and which use a trial-and-error strategy, guessing the number of cookies for each child and adjusting until the number is right. Using physical models to emphasize that in some contexts remainders can be cut into fractional pieces but in others they cannot can help students distinguish different types of division contexts.

Knowledge of assessment

Assessment tasks, like instructional tasks, should routinely engage students in creating and interpreting diagrams. Do the tasks used in assessments in your school or district allow you opportunities to understand the ways in which your students think about multiplicative situations? Do these tasks focus on the big ideas related to multiplication and division? When do your students interpret diagrams? When do they create their own diagrams? How do their responses to the tasks help you craft your next instructional moves? Moreover, do your students have opportunities to create problem situations, writing word problems like those displayed in figures 3.6–3.9? Using this assessment strategy can provide you with an opportunity to assess your students' understanding of critical features of division situations.

Conclusion

The ideas and samples of student work in this chapter have illustrated the complexity and challenges that you face in teaching essential understandings related to division. If your students are to be successful in developing these understandings, you must not minimize the complexity of division by trying to integrate all of its components in a few superficial lessons. Instead, you must emphasize different problem types, models, representations, contexts, and strategies as your students build their understanding of the meaning of division. Supporting them in this work requires carefully selecting tasks and posing effective questions that will lay a foundation for them to develop essential understandings related to the properties of multiplication and division—the subject of the next chapter.

practice

Chapter 4
Multiplication and Division Properties

Big Idea 2

The properties of multiplication and addition provide the mathematical foundation for understanding computational procedures for multiplication and division, including mental computation and estimation strategies, invented algorithms, and standard algorithms.

Essential Understanding 2*a*

The commutative and associative properties of multiplication and the distributive property of multiplication over addition ensure flexibility in computations with whole numbers and provide justifications for sequences of computations with them.

Essential Understanding 2*b*

The right distributive property of division over addition allows computing flexibly and justifying computations with whole numbers, but there is no left distributive property of division over addition and no commutative or associative property of division of whole numbers.

Essential Understanding 2*c*

Order of operations is a set of conventions that eliminates ambiguity in, and multiple values for, numerical expressions involving multiple operations.

Big Idea 2 in *Developing Essential Understanding of Multiplication and Division for Teaching Mathematics in Grades 3–5* (Otto et al. 2011) focuses on properties as the foundation for computing with multiplication and division. Developing students' understanding of these properties is the subject of this chapter. Essential Understandings 2*a*, 2*b*, and 2*c*—the first three essential understandings that Otto and colleagues associate with Big Idea 2—single out and highlight particular properties that ensure flexibility and guard against ambiguity in these computations. This chapter investigates approaches for moving students toward Big Idea 2—especially by helping them grasp these three essential understandings related to individual properties of multiplication and division. The fourth and final essential understanding that Otto and colleagues associate with Big Idea 2 focuses on algorithms for multiplying and dividing, and these are the subject of Chapter 5.

Working toward Big Idea 2 through Essential Understandings 2*a*, 2*b*, and 2*c*

Recognizing and applying mathematical properties is an important part of students' mathematical activity. These properties can be useful for simplifying calculations, performing mental computations, and recognizing expressions that are mathematically equivalent. However, it is just as important that students think about the mathematical ideas that underlie these properties, justify their application, and consider domains where they do not apply. In many cases, teachers may find it difficult to facilitate their students' understanding of the properties, since they themselves were simply told that they were true without receiving much explanation. Further, although teachers may recognize the properties in mathematical situations, they may not be obvious to students.

As Otto and colleagues (2011) emphasize in their discussion of Big Idea 2, it is important that teachers realize that these properties are truly powerful ideas. They are not simply terms and patterns to be memorized, but tools to be used to solve problems and support mathematical reasoning. Because the properties of multiplication and division are not necessarily obvious to students, you must help them come to understand them and use them as tools to assist in computing and reasoning.

Figure 4.1 presents several tasks that problem solvers can simplify by applying different properties of multiplication and division. Examine these tasks while responding to the questions in Reflect 4.1. For this work, regard the names of the properties as less important than the ideas that they represent.

Reflect 4.1

What ideas related to multiplication and division are associated with the tasks shown in figure 4.1?

How do these ideas come into play in thinking about and solving these tasks?

- Seth has 37 boxes of marbles with 28 marbles in each box. Marcus has 28 boxes of marbles with 37 marbles in each box. Who has more marbles?
- I am thinking of two whole numbers that when multiplied together give a product of 0. What can you tell me about the two whole numbers?
- 50 × 40 is 2,000. What is an easy way to find 49 × 40?
- What is 5 + 5 + 5 × 0?
- Which is more: 6 ÷ 0 or 0 ÷ 6?
- Determine the missing numbers that make these statements true:

  18 × 6 is the same as ___ × 12   24 × 12 is the same as 6 × ___

  200 ÷ 6 is the same as 100 ÷ ___   200 ÷ 6 is the same as ___ ÷ 12

Fig. 4.1. Problems that can be solved efficiently by applying properties of multiplication and division

This chapter addresses the tasks in figure 4.1 throughout and discusses how students in grade 3–5 approached them. However, the chapter begins with an examination of some common misconceptions that students demonstrate in relation to multiplication and division.

Addressing misconceptions about multiplication and division

Students attend to many different aspects of problems when learning mathematics. Often, those aspects are relevant to the situation, but sometimes students focus on an irrelevant feature and assume that it is actually important. In this way, they may create or reinforce a misconception. Consider the task and samples of student work shown in figure 4.2, and respond to the questions in Reflect 4.2.

Reflect 4.2

Figure 4.2 shows the work of three students at different grade levels in response to a claim by a fictitious student, Kara, that "when you multiply two numbers, the answer is always bigger."

What misconception frequently held by students is this task designed to bring to the surface and address?

Reflect 4.2. *Continued*

Which of the three students whose work is shown in the figure appear to hold this misconception?

How would you challenge this misconception and help students move forward?

Megan's (grade 3) explanation of her disagreement with Kara's claim

> 7. Kara says that when you multiply two numbers the answer is always bigger.
>
> Do you agree with Kara? Circle one: Yes   (No)
>
> Explain your thinking. If I was multiplying 4×0 the ansrer would be 0 so it is sometimes bigger and some times it would be smaller.

Jason's (grade 4) explanation of his agreement with Kara's claim

> 3. Kara says that when you multiply two numbers the answer is always bigger.
>
> Do you agree with Kara? Circle one: (Yes)   No
>
> Explain your thinking.
>
> because when you multiply it is always higher ant when you divide it is smaller

Fig. 4.2. Three students' agreement or disagreement with Kara's claim

Karen's (grade 5) explanation of her agreement with Kara's claim

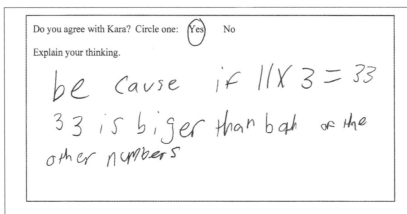

Do you agree with Kara? Circle one: (Yes)   No

Explain your thinking.

$$be \ cause \ if \ 11 \times 3 = 33$$
$$33 \ is \ biger \ than \ both \ of \ the$$
$$other \ numbers$$

Fig. 4.2. *Continued*

We gave this task to 175 students in the third, fourth, and fifth grades to see whether they held the misconception that "multiplication makes bigger" (Greer 1994). Nearly half of the students (85, or 49 percent) agreed with Kara. Many students, like Karen, the fifth grader whose work is shown in figure 4.2, provided a specific example of a product that was greater than either of the numbers multiplied to obtain it. Others made general statements, referring to addition or skip counting, and stating that addition makes numbers larger. As students progress in their learning, they tend to develop a sense that adding numbers always results in a larger number and subtracting results in a smaller number. Although this is not always true, it is true when the numbers being added are positive whole numbers. In the same way, in multiplication, whenever both factors are greater than 1, the product will be greater than either factor. Because elementary students work primarily with positive integers, they may assume that this phenomenon is a property of multiplication.

The 86 students (49 percent) who disagreed with Kara cited multiplying by 1 or by 0, as did Megan, the third grader whose work appears in figure 4.2. These students provided an appropriate counterexample and do not appear to hold the misconception that "multiplication makes bigger." However, we do not know if they would also recognize that multiplying by a positive rational number less than 1—for example, $\frac{1}{2}$ or 0.98—results in a product that is less than one of the factors.

Figure 4.3 displays three problems designed to bring to the surface misconceptions that students commonly form about multiplication and division. Reflect 4.3 offers guidance for considering the misconceptions that may surface when students approach these problems.

Reflect 4.3

Write an equation for each of the problems shown in figure 4.3.

What specific misconceptions about multiplication and division might these questions address?

- Jamil has 4 dollars. He wants to buy some games that cost 20 dollars each. How many games can he buy?

- Dora spent 4 dollars and bought 20 pencils. How much did each pencil cost?

- Zena has 2 cups of cheese to make miniature pizzas. She will put 1/3 of a cup of cheese on each miniature pizza. How many miniature pizzas can she make?

Fig. 4.3. Problems designed to bring students' misconceptions about multiplication and division to the surface

A common misconception that students hold about division is that the dividend must be greater than the divisor. When Rod, a third-grade student, was asked about the expression $4 \div 20$, he exclaimed, "$4 \div 20$ doesn't make any sense! How would you divide 4 things into 20 groups?" In fact, in some contextual situations, $4 \div 20$ does not make much sense. In the first scenario in figure 4.3, for instance, Jamil can buy 0 games, unless it is possible to buy $^4/_{20}$ of a game. Working with problems of other types, set in other contexts, may help students clarify the idea that the expression $4 \div 20$ does have meaning and that the quotient will be less than 1. For example, in the second scenario in figure 4.3, Dora spent a total of $4 on 20 pencils, so each pencil cost $4 \div 20$, or $^4/_{20}$ of a dollar, or 20 cents.

Another common misconception relates to the two previously mentioned. In addition to agreeing with the fictitious student Kara that "multiplying makes bigger," Jason, whose thinking is displayed in figure 4.2, also appears to hold the misconception that "division makes smaller" (Greer 1994). Students are likely to make this inference because the quotient is smaller than the dividend in nearly every division statement they see with whole numbers—that is, when the dividend is greater than the divisor. This would not be the case, however, if the divisor is a positive rational number less than 1. For example, the equation $2 \div {}^{1}/_{3} = 6$ fits the third scenario in figure 4.3, with Zena's miniature pizzas, and has a quotient that is greater than both the dividend and the divisor.

Developing understanding of the commutative property

Students begin developing an understanding of the idea of commutativity in their early work with addition. Translating this idea to multiplication is part of developing an understanding of Big Idea 2—in particular, Essential Understanding 2*a*—and becoming skilled in using multiplication in problem solving.

To begin exploring ways of assisting students in building an understanding of the commutative property of multiplication, consider situations A and B in figure 4.4, using the questions in Reflect 4.4 to guide your examination.

Reflect 4.4

Which of the situations in figure 4.4, A or B, would you expect to be more difficult for your students?

What connections would you expect your students to make between these two situations?

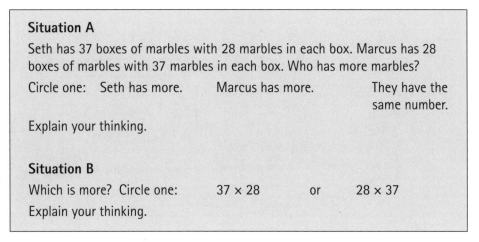

Situation A

Seth has 37 boxes of marbles with 28 marbles in each box. Marcus has 28 boxes of marbles with 37 marbles in each box. Who has more marbles?

Circle one: Seth has more.   Marcus has more.   They have the
                                                                same number.

Explain your thinking.

Situation B

Which is more? Circle one: 37 × 28   or   28 × 37

Explain your thinking.

Fig. 4.4. Two situations involving a comparison of 37 × 28 and 28 × 37

To adults, it may seem obvious that the quantities being compared in both situations, 37 × 28 and 28 × 37, are the same; their equality is guaranteed by the commutative property of multiplication—$a \times b = b \times a$ for all real numbers a and b—hereafter referred to simply as the *commutative property*. However, for students who are unfamiliar with this property, these situations are much more challenging.

When we gave the situations shown in figure 4.4 to fourth- and fifth-grade students, most recognized that the quantities being compared in each situation are equal. Ninety percent of the fourth-grade students determined that the results for Seth and Marcus were the same in situation A, and 95 percent said that the results of the multiplications in situation B were the same. Of the fifth-grade students, 96 percent recognized that the results were the same in situation A, and 94 percent said that the results in situation B were the same. Yet, the ways in which students made this determination and explained their reasoning varied. Use the questions in Reflect 4.5 to aid in evaluating the samples of student explanations displayed in figures 4.5–4.8.

Reflect 4.5

Figures 4.5–4.8 show the responses and corresponding explanations offered by four fifth-grade students to situation A or situation B, displayed in figure 4.4.

Which students' explanations would you consider strong?

What questions do these examples raise about these students' understanding?

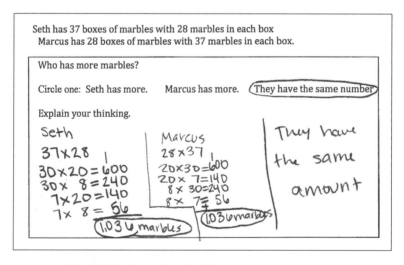

Fig. 4.5. Andrew's (grade 5) response to situation A

Seth has 37 boxes of marbles with 28 marbles in each box
 Marcus has 28 boxes of marbles with 37 marbles in each box.

Who has more marbles?

Circle one: ~~Seth has more.~~   Marcus has more.   They have the same number

Explain your thinking. (Seth)   (Marcus)

$$
\begin{array}{cc}
37 & 28 \\
\times 28 & \times 37 \\
\hline
296 & 196
\end{array}
$$

Fig. 4.6. Bart's (grade 5) response to situation A

Seth has <u>37 boxes</u> of marbles with <u>28 marbles</u> in each box
 Marcus has 28 boxes of marbles with <u>37 marbles</u> in each box.

Who has more marbles?

Circle one: Seth has more.   Marcus has more.   (They have the same number)

Explain your thinking.

because Seth has more boxes but less
marbles, and marcas has less boxes but
  in each box   more marbles in
                       each box

Fig. 4.7. Cassandra's (grade 5) response to situation A

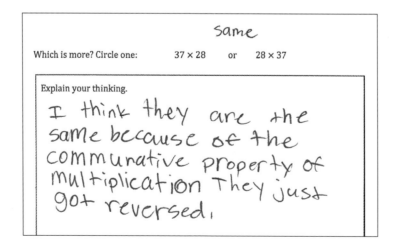

same

Which is more? Circle one: 37 × 28   or   28 × 37

Explain your thinking.

I think they are the
same because of the
communative property of
multiplication They just
got reversed.

Fig. 4.8. Deron's (grade 5) response to situation B

Andrew, whose work is displayed in figure 4.5, recognized that in situation A Seth and Marcus have the same number of marbles, and he provided calculations to support his conclusion. Bart, whose work appears in figure 4.6, also provided calculations, though both are incorrect, and he stated that Seth has more marbles. These two examples demonstrate that it is not obvious to all students—even fifth graders—that the two quantities of marbles in situation A are the same. Students may recognize and apply the commutative property correctly in some situations—for example, when presented with symbolic representations—but not in other situations, as when working with contextual situations. It is critical to help students recognize and apply this and other properties in a variety of situations. Asking students to solve problems in a contextual situation such as situation A and in a symbolic context such as situation B and then having them discuss the related ideas can help them develop these essential understandings.

It is also important to ensure that your students' understanding of the commutative property includes *generality*. Do your students recognize that this property applies to *any* two whole numbers? Schifter and colleagues (2008) demonstrate that some students recognize and apply a property to relatively small, familiar numbers but are unsure whether the property applies to larger, less familiar ones. It is critical that students understand how to model the commutative property in a way that explains or justifies it for any two whole numbers. Consider the questions in Reflect 4.6.

Reflect 4.6

How do you know that 7×4 and 4×7 generate the same result?

How do you know that the result will be the same for any two numbers that you multiply?

Are there situations in which you multiply but cannot "switch the numbers around" to generate the same result?

You may know that 7×4 and 4×7 are the same because you know both number facts: $7 \times 4 = 28$ and $4 \times 7 = 28$. How does the equality of these two facts relate to other facts? What underlying mathematical relationship exists that ensures that the commutative property applies to any two whole numbers? Again, remember that it is not always obvious to students that having 45 groups of 78 objects is the same as having 78 groups of 45 objects.

An array can be a powerful model to help students recognize and extend the commutative property to other numbers. Consider the array displayed in figure 4.9.

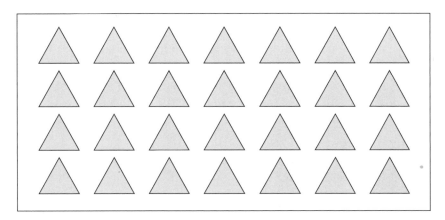

Fig. 4.9. A 4-by-7 array

It is easy to see 4 groups of 7 objects by considering each row as a group. Is there also a way to see 7 groups of 4 objects? Yes, considering each column as a group gives 7 columns with 4 objects in each column. To what extent does this relationship apply to other values for the number of rows and columns?

You could have students create arrays for various values until they recognize and state that this relationship applies to any situation, although students in grades 3–5 are likely to base their conclusions on work with whole numbers. Note that this general relationship occurs in any array (and can be extended to situations beyond the whole numbers through use of a continuous model, such as a rectangular array formed by lengths of line segments). This justification of the commutative property is not based only on a few examples, but also on the fact that in any array, the groups can be seen as the rows or the columns.

Returning to this idea periodically with your students can help you assess whether they recognize the general nature of the commutative property and can justify it on the basis of the general relationship that exists between the number of multiplicative units and the size of each multiplicative unit. Further, you will want to ensure that your students recognize and are able to apply commutativity to multiplication and understand that commutativity does not apply to division, except in cases where a number is divided by itself, such as $30 \div 30$. Students should think about situations that involve $20 \div 5$ and $5 \div 20$ and be able to explain why these two expressions are not equivalent and extend this to other instances with division.

Furthermore, multiplication is not always commutative. For example, you may recall from your work with matrices in high school that matrix multiplication is not commutative. It is important for students to ponder and explain when and why commutativity can be applied rather than simply memorize a rule.

Commutavity is a very powerful property that reduces and simplifies calculations. For example, knowing that multiplication is commutative means that knowing 5×8 also means knowing 8×5. Moreover, it simplifies calculations, enabling students to determine 5×8 by thinking about 8 groups of 5 and recognizing that two 5s make 10, so eight 5s are the same as four 10s and make 40. This is easier than trying to determine how many are in 5 groups of 8.

It is important to consider the textbook materials that you have available and the extent to which they promote a deep understanding of the commutative property. In many textbooks that we examined, the commutative property was simply stated as true or "proved" true for students by having them looking at 3 groups of 4 objects and noticing that it has the same total as 4 groups of 3 objects, instead of helping them develop a deeper understanding of the relationship between the groups through the use of arrays, as discussed earlier.

Furthermore, many textbooks state the commutative property and then encourage students to "fill in the blanks" for situations such as $6 \times 8 = 8 \times$ ___ . Such tasks do not permit distinguishing between students who have a superficial understanding that enables them to fill in the blanks correctly with missing numbers and those who have a deeper understanding of the commutative property of multiplication.

We suggest introducing various properties through situations that invite students to think about repackaging objects, such as the situations in tasks 1 and 2 in figure 4.10. Read through the tasks in the figure, guided by the questions in Reflect 4.7.

Reflect 4.7

Which ideas would you expect to come to the surface if your students were given the tasks in figure 4.10?

What strategies would you expect your students to use in working on these tasks?

1. How many different ways can you group 24 objects so that the same number of objects is in each group?

2. You have 6 packages of pencils, with 4 pencils in each package. How many packages can you make if you want to put 6 pencils in each package?

3. Fill in the blank to make the equation true: 6 × 4 = ___ × 6.

Fig. 4.10. Three tasks that invite students to think about ideas related to commutativity

You might use task 1 to introduce students to various ideas related to grouping objects. Furthermore, it can allow you to assess whether your students recognize that if they can make 3 groups of 8, then they can also make 8 groups of 3. Regrouping quantities in this way is not something that is obvious to students to do, so many students in third grade, and perhaps even students who have been introduced to the commutative property of multiplication more formally, may not use this idea. Comparing and contrasting tasks like tasks 2 and 3 can help students see a connection between a contextual situation and a corresponding symbolic representation of the commutative property. Providing situations like these periodically will encourage your students to continue making these connections. Furthermore, throughout the year, it is important to discuss situations in which using the commutative property can make calculations easier. Revisiting this property can help your students retain their understanding and recognize the power of this property.

Developing understanding of the distributive property

The distributive property of multiplication over addition, referred to hereafter simply as the *distributive property*, is one of the most powerful and important properties related to multiplication and division. It provides a tool for simplifying calculations and undergirds all multi-digit algorithms for multiplication. Despite its central role in computation with multiplication, students may not recognize or apply the distributive property in a variety of situations. Consider the array in figure 4.11, with 6 rows and 8 columns—that is, a 6-by-8 array.

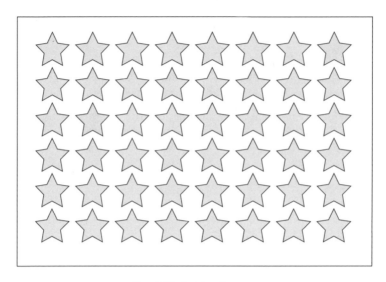

Fig. 4.11. A 6-by-8 array

This array could be viewed as 6 groups of 8 or, variously, as 3 groups of 8 and 3 more groups of 8, 5 groups of 8 and 1 more group of 8, or 2 groups of 8 and 4 more groups of 8. Examining a model such as this promotes the idea of the distributive property and helps students forge connections with it as they consider that $6 \times 8 = 3 \times 8 + 3 \times 8 = 5 \times 8 + 1 \times 8 = 2 \times 8 + 4 \times 8$.

This property simplifies computations and quick recall of multiplication facts. Students who do not yet have immediate recall of 6×8 can quickly determine that 5×8 is 40 and then add another 8 for the sixth group of 8, yielding a total of 48. This property also assists students in performing multi-digit computation, since, for example, they can determine 6×12 by calculating 6×10 and adding 6×2. Such strategies are often natural ways that students reason before they have been formally introduced to the distributive property.

Before considering samples of student work, take time to think about possibilities for using the distributive property to simplify calculations. Examine the multiplication problems in Reflect 4.8, working with colleagues if possible.

Reflect 4.8

You know that 40 × 40 = 1,600. Use this information to compute the following:

   (a) 40 × 39 = ____

   (b) 40 × 41 = ____

   (c) 41 × 40 = ____

   (d) 39 × 41 = ____

What strategies did you use, along with the fact that 40 × 40 = 1,600, to help you arrive at the product in each case?

Applying the distributive property to these related facts provides a powerful tool for determining reasonable and correct answers efficiently—more quickly than would be possible by using other methods. For (a), 40 × 39, you could visualize 40 groups of 39 and recognize that 40 groups of 40 (that is, 40 × 40) has one more object in each group, or 40 more objects. Therefore, 40 × 39 = 40 × 40 – 40 × 1 = 1,560. For (b), 40 × 41, you might realize that you have the same number of groups as in 40 × 40, but now you have one more object in each group. Therefore, 40 × 41 = 40 × 40 + 40 × 1 = 1,640. For (c), 41 × 40, you might recognize that this product is the same as for (b), or you might realize that now you have one more group of 40 than in 40 × 40, since 41 × 40 means that you have 41 groups of 40 rather than 40 groups of 40. Therefore, 41 × 40 = 40 × 40 + 1 × 40 = 1,640. You could compute the product in (d), 39 × 41, by considering the product in (b), 40 × 41 = 1,640, noting that in (d) you have one fewer group of 41. Therefore, 39 × 41 = 40 × 41 – 1 × 41 = 1,599. Notice how the distributive property facilitates relating facts and recognizing the relationships among these facts on the basis of the meaning of multiplication. (You might want to extend your reasoning in (d) to consider how 20 × 20 relates to 19 × 21, 12 × 12 relates to 11 × 13, and 25 × 25 relates to 24 × 26, for example. Sketch arrays and think about this relationship.)

To gain insight into students' understanding of the distributive property, we provided the task shown in figure 4.12 to fourth- and fifth-grade students. Examine the task, and consider the questions in Reflect 4.9.

Reflect 4.9

How does the following problem relate to the distributive property?

What strategies would you expect your students to use in this situation?

---

50 × 40 is 2,000. What is an easy way to find 49 × 40?

Explain your easy way.

Fig. 4.12. A task intended to reveal students' understanding of the use of the distributive property

You could visualize 50 × 40 as 50 groups of 40 and 49 × 40 as 49 groups of 40. To determine the result for 49 × 40, you could subtract 40 from 50 × 40 because 50 × 40 = 49 × 40 + 1 × 40. Therefore, 50 × 40 – 1 × 40 = 49 × 40. As in the case of the 6-by-8 array displayed in figure 4.11, recognizing this relationship involves applying the meaning of multiplication with 40 as the multiplicative unit and determining the difference between having 49 groups of the multiplicative unit and having 50 groups of the multiplicative unit.

What strategies did the fourth- and fifth-grade students in our group use to approach this situation? Very few students—only about 8 percent of the fourth graders and 3 percent of the fifth graders—calculated 49 × 40 by subtracting 40 from 50 × 40. The students did not seem to consider or apply the meaning of multiplication (for 50 groups of 40) or of the distributive property. Instead, what strategies did these students use? Figures 4.13–4.15 show samples of common strategies offered as "easy ways" to calculate 49 × 40.

50 × 40 is 2000. What is an easy way to find 49 × 40? Explain your easy way.

My easy way to find 49 × 40. I think stackeing is because you get the answer faster.

$$\begin{array}{r} 49 \\ \times\ 40 \\ \hline 160 \end{array}$$

Fig. 4.13. Michael's (grade 4) "easy" strategy for 49 × 40

50 × 40 is 2000. What is an easy way to find 49 × 40? Explain your easy way.

My easy way to find 49 × 40.

Find out what is 50 less than 50X40.

Fig. 4.14. Luke's (grade 4) "easy" strategy for 49 × 40

50 × 40 is 2000. What is an easy way to find 49 × 40? Explain your easy way.

My easy way to find 49 × 40.

My easy way to find 49x40 is to multiply traditionaly. (A = 1,960)

$$\begin{array}{r} {}^{3}49 \\ \times\ 40 \\ \hline 1960 \\ \hline 1,960 \end{array}$$

Fig. 4.15. Timothy's (grade 5) "easy" strategy for 49 × 40

For the fourth-grade students in our sample, the most common approach was to perform, incorrectly, a procedure for multiplying 49 × 40. Michael, the fourth grader whose work is shown in figure 4.13, appears to have multiplied 9 by 0 and 4 by 4, generating a result of 160. Luke, the fourth grader whose work is shown in figure 4.14, attempted to use a related fact, but suggested subtracting 50 rather than 40 from 2,000. The most common approach for fifth-grade students in our sample was similar to that of Timothy, the fifth grader whose work is displayed in figure 4.15. Students at this grade level frequently used a standard algorithm to determine 49 × 40. The next chapter discusses the teaching and learning of multiplication and division algorithms; what is noteworthy in the context of this chapter's discussion is that many students in these classrooms relied solely on the traditional algorithm or incorrectly generated a procedure for 49 × 40 instead of considering the meaning of multiplication, the relationship between 50 × 40 and 49 × 40, and the distributive property.

When you examine the tasks in your textbook, consider the nature of the situations that they involve. Students naturally apply the distributive property in situations that involve repeated doubling (for example, 4 × 7 = 2 × 7 + 2 × 7), and the tasks in various textbooks reinforce this idea. However, students find it much more challenging to write expressions using formal notation, since many are unfamiliar with expressions or equations with more than one operation. Using arrays to create pictorial models and building toward the formal notation, as illustrated in figure 4.16, can help students make these connections. The textbooks that we looked at were quick to introduce formal notation to students and have them "fill in the blanks," writing equations such as 4 × 9 = 4 × 5 + 4 × ___ = ____, for example. Such notation may confuse students and move the focus away from the meaning behind the symbols. By contrast, connecting symbolic and pictorial representations can help students develop an understanding of the meaning of the distributive property and the formal symbols.

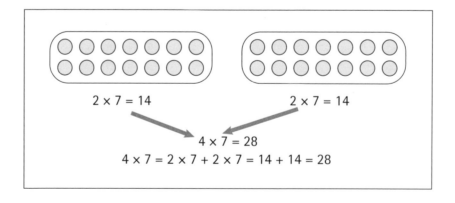

Fig. 4.16. Connecting representations for 4 × 7

Consider using some of the problems in figure 4.17 to introduce and formalize your students' thinking about the distributive property. After allowing students to work individually on these situations and then having them share approaches with the class, take the opportunity to introduce the formal name, *distributive property*.

1. Marsha has 7 bags of marshmallows with 5 marshmallows in each bag. Ryan has 9 bags of marshmallows with 5 marshmallows in each bag. Who has more marshmallows? How many more marshmallows does that person have?

2. Marsha has 7 bags of marshmallows with 6 marshmallows in each bag. Ryan has 7 bags of marshmallows with 8 marshmallows in each bag. Who has more marshmallows? How many more marshmallows does that person have?

3. Marsha has 6 bags of marshmallows with 10 marshmallows in each bag. Ryan has 6 bags of marshmallows with 9 marshmallows in each bag. How many marshmallows does Marsha have? How many marshmallows does Ryan have?

4. Marsha has 10 bags of marshmallows with 5 marshmallows in each bag. Ryan has 13 bags of marshmallows with 5 marshmallows in each bag. How many marshmallows does Marsha have? How many marshmallows does Ryan have?

Fig. 4.17. Samples of useful problem situations for introducing the distributive property

Problem 1 can help students recognize how much more 9×5 is than 7×5 by encouraging them to consider what happens when the number of multiplicative units is increased by 2: Ryan and Marsha both have bags of marshmallows with 5 marshmallows in a bag, but Ryan has 2 more bags than Marsha. Problem 2 keeps the number of multiplicative units—bags—the same and increases the size of the multiplicative unit. Students should recognize that Ryan has 2 more marshmallows in each of his 7 bags, or a total of 14 more marshmallows. Problem 3 encourages students to recognize that if they know 6×10, then they can determine 6×9 by subtracting 6, since each of Ryan's 6 bags has one fewer marshmallow than each of Marsha's 6 bags. In problem 4, Marsha and Ryan again have bags of marshmallows with 5 marshmallows to a bag, as in problem 1, but this time Ryan has 13 bags, or 3 more bags than Marsha. This problem can help students forge a connection with two-digit by one-digit multiplication as they link 10×5 with 13×5 by identifying it as 3×5 more. Using a mixture of situations such as those in problems 1 and 2, and then moving toward the use of situations similar to those in problems 3 and 4, can assist students in recognizing the power of the distributive property and applying it for efficient problem solving.

Developing understanding of the associative property

As in the case of the commutative and distributive properties, textbooks for grades 3–5 may present the associative property of multiplication. Often they introduce this property, hereafter referred to as the *associative property*, in the manner shown in figure 4.18.

Factors in multiplication can be grouped together in different ways, and the product is always the same.

Example:

$(4 \times 3) \times 2 = 4 \times (3 \times 2)$

$\quad 12 \quad \times 2 = 4 \times \quad 6$

$\quad\quad\quad 24 = 24$

Complete each equation.

$4 \times (8 \times 3) = (\underline{\quad} \times 8) \times 3$ $\qquad\qquad$ $5 \times (4 \times 6) = (5 \times \underline{\quad}) \times 6$

Use parentheses to show two ways of grouping. Find the product.

$5 \times 2 \times 4 =$ $\qquad\qquad\qquad\qquad$ $7 \times 6 \times 5 =$

Fig. 4.18. A presentation of the associative property
similar to that found in many textbooks

Although this presentation is mathematically correct, it does not provide students with a compelling reason for using the associative property. It also does not show how this property might be used in a context outside of school. Furthermore, it requires students to view an instance of the associative property with particular numbers (such as 4, 3, and 2) and make a generalization that this property applies to all numbers. As we have stated before, these properties may not be obvious to students.

We used two different types of tasks—"missing number" and "repackaging" tasks—with 170 students in grades 4 and 5. Figures 4.19 and 4.20 show our groups of tasks. Read through these tasks, examining them through the lens of the questions in Reflect 4.10.

Reflect 4.10

What are some different ways in which students might solve the missing number tasks in figure 4.19?

Compare missing number tasks A and B in figure 4.19 with the repackaging tasks in figure 4.20. How might students solve these tasks differently?

Determine the missing numbers in the following tasks. Explain your thinking for each part.

    A. 18 × 6 is the same as ___ × 12

    B. 24 × 12 is the same as 6 × ___

    C. 200 ÷ 6 is the same as 100 ÷ ___

    D. 200 ÷ 6 is the same as ___ ÷ 12

Fig. 4.19. Missing number tasks

1. Rosa has 18 bags with 6 marbles in each bag. She wants to repackage the marbles with 12 marbles in each bag. How many bags will Rosa need?

2. Lucy has 24 bags with 12 marbles in each bag. She wants to repackage the marbles so that she uses 6 bags and she has the same number of marbles in each bag. How many marbles will Lucy have in each bag?

Fig. 4.20. Repackaging tasks

Students who were successful in missing number tasks A and B were also able to solve the repackaging tasks, and they usually used the same method to complete both types of tasks. However, there were a few exceptions. Consider the work of two fifth-grade students, Dalton and Kelsey, shown in figures 4.21 and 4.22, respectively.

2. Determine the missing number: 18 × 6 is the same as ___9___ × 12? Explain your thinking.

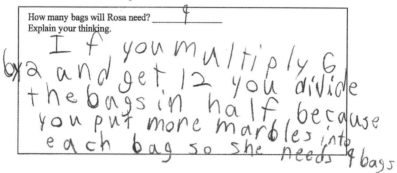

I did 6×10 and 6×8 and added the products together and got 108 which is also 9×12.

8. Rosa has 18 bags with 6 marbles in each bag. She wants to repackage the marbles with 12 marbles in each bag.

How many bags will Rosa need? ___9___
Explain your thinking.

If you multiply 6 by 2 and get 12 you divide the bags in half because you put more marbles into each bag so she needs 9 bags

Fig. 4.21. Dalton's (grade 5) work on related missing number and repackaging tasks

2. Determine the missing number: 18 × 6 is the same as ___9___ × 12? Explain your thinking.

I know that 6×2=12 and so it would be 18×12 = ___ × 12 and since I did 6×2 I have to do 18÷2 to make both sides even and that equals 9!

8. Rosa has 18 bags with 6 marbles in each bag. She wants to repackage the marbles with 12 marbles in each bag.

How many bags will Rosa need? _9 bags_   12 ÷ 9 = 108
Explain your thinking.

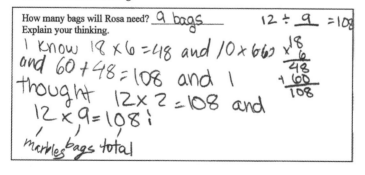

I know 18×6=48 and 10×6=60
and 60+48=108 and I
thought 12×2=108 and
12×9=108!
marbles bags total

Fig. 4.22. Kelsey's (grade 5) work on related missing number and repackaging tasks

The samples of work from Dalton and Kelsey both illustrate the use of direct computation (which was the method most commonly used on these tasks) as well as the idea of taking advantage of a doubling-and-halving strategy. The doubling-and-halving strategy is a special case of the associative property. For example, $18 \times 6 = (9 \times 2) \times 6 = 9 \times (2 \times 6) = 9 \times 12$. For missing number task B, some students noted that 6 is half of 12 and used doubling and halving. Students who use this strategy are employing both the associative and the commutative properties. Our group of students included a few students who completed only the repackaging task, providing evidence that contextualized tasks can be more accessible to students than decontextualized tasks in some cases.

Students in our group had a more difficult time with missing number tasks C and D in figure 4.19, and very few fourth- or fifth-grade students responded to task D correctly. Analyze the work of Isabel, Max, Daniella, and Norris, shown in figures 4.23–4.26, using Reflect 4.11 as a guide.

Reflect 4.11

Figures 4.23, 4.24, 4.25, and 4.26 show, respectively, the work of Isabel, Max, Daniella, and Norris on either missing number task C or missing number task D, shown in figure 4.19.

On the basis of their work, what do you think Isabel, Max, Daniella, and Norris appear to understand about multiplication and division?

Using the repackaging tasks in figure 4.20 as models, write a contextual item that might help students understand missing number task C in figure 4.19.

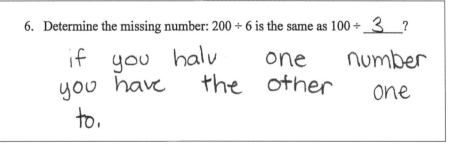

6. Determine the missing number: $200 \div 6$ is the same as $100 \div$ __3__ ?

> if you halv one number
> you have the other one
> to.

Fig. 4.23. Isabel's (grade 5) work on the missing number task C

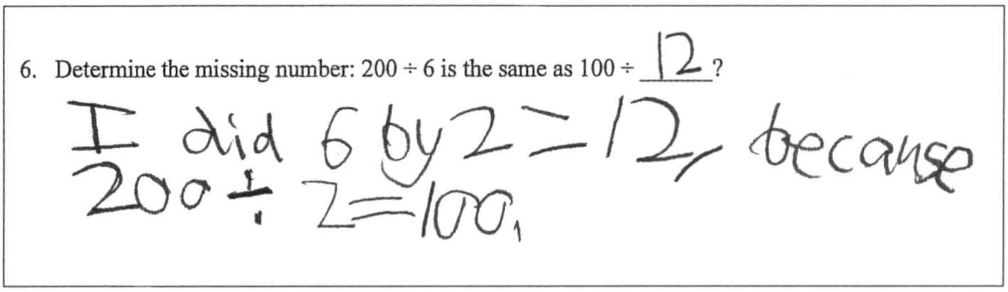

6. Determine the missing number: 200 ÷ 6 is the same as 100 ÷ 12 ?

I did 6 by 2 = 12, becanse 200 ÷ 2 = 100,

Fig. 4.24. Max's (grade 4) work on missing number task C

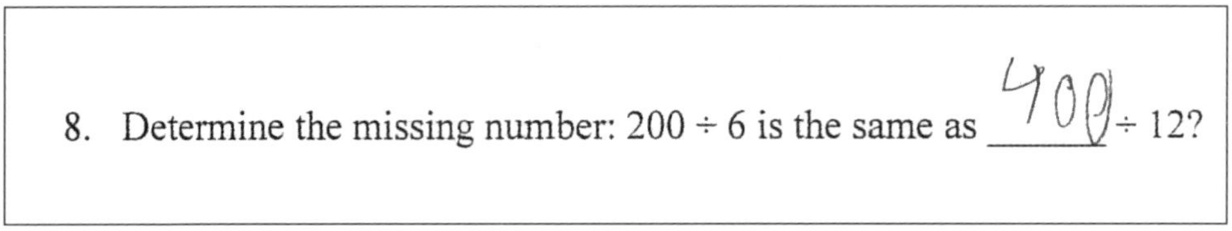

8. Determine the missing number: 200 ÷ 6 is the same as 400 ÷ 12?

Fig. 4.25. Daniella's (grade 4) work on missing number task D

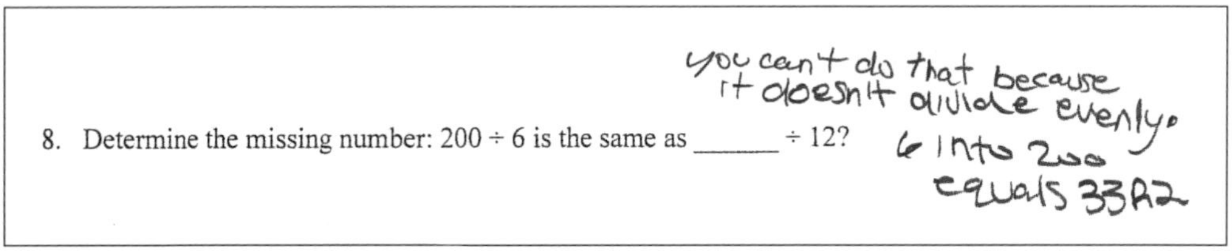

8. Determine the missing number: 200 ÷ 6 is the same as _____ ÷ 12? you can't do that because it doesn't divide evenly. 6 into 200 equals 33A2

Fig. 4.26. Norris's (grade 5) work on missing number task D

Of the four students whose work is shown, Isabel (fig. 4.23) appears to have the clearest understanding of the fact that the quotient will remain the same when the dividend and divisor are multiplied by the same factor. Daniella (fig. 4.25) may also have this understanding, but she did not provide any explanation for her answer. This property is used in mental arithmetic, by recognizing, for example, that $2800 ÷ 700 = 28 ÷ 7$, and that $300 ÷ 75 = 100 ÷ 25$, and $10 ÷ 2.5 = 40 ÷ 10$.

Students who understand that division can be represented as a fraction may be better equipped to complete these tasks successfully. Missing number task D could be rewritten as $200/6$ is the same as $?/12$. None of the students in our sample actually did this, and very few students completed missing number tasks C and D correctly, as the table in figure 4.27 shows.

Missing number task C 200 ÷ 6 is the same as 100 ÷ ___		Missing number task D 200 ÷ 6 is the same as ___ ÷ 12	
Grade 4 (n = 19)	Grade 5 (n = 21)	Grade 4 (n = 60)	Grade 5 (n = 62)
5% correct (1 of 19)	43% correct (9 of 21)	12% correct (7 of 60)	15% correct (9 of 62)

Fig. 4.27. Percentages (and numbers) of fourth- and fifth-grade students who correctly answered missing number tasks C and D

Overall, students approached missing number tasks C and D differently. In task C, 10 out of 19 fourth-grade students and 3 out of 21 fifth-grade students (53 percent and 14 percent, respectively) used a doubling-and-halving strategy, in the same manner as Max (see fig. 4.24). Some fifth-grade students attempted to calculate the quotient directly (4 out of 21, or 19 percent) but were unable to move forward from that point.

Students typically have more difficulty solving proportions similar to that in task D than that in task C. In the case of task D, where the unknown is in the upper position of one of the ratios, the most common response from fourth-grade students was to leave the item blank (55%, or 33 out of 60). Students in the fifth grade typically attempted a direct calculation and arrived at an incorrect answer (10 percent, or 6 out of 62), or they calculated correctly and then reasoned that the problem could not be solved (13 percent, or 8 out of 62), as illustrated by Norris's work in figure 4.26.

It may be helpful to rewrite these missing number tasks in contextual situations, as in figure 4.28. Students are often more successful when they are working on tasks of this type within a context as opposed to solving decontextualized tasks such as those shown in figure 4.19.

- Hank has 200 feet of string that has been cut into 6 pieces of equal length. Ken has 100 feet of string cut into pieces that are the same length as Hank's pieces. How many pieces does Ken have?

- Hank has 200 feet of string that has been cut into 6 pieces of equal length. Mia has 12 pieces of string that are the same length as Hank's pieces. How long was Mia's string before it was cut?

Fig. 4.28. Cutting string tasks (contextualized missing number tasks)

Developing understanding of multiplication and division with 1 and 0

Mathematics textbooks for grades 3–5 often introduce properties of multiplication and division related to 1 and 0, and these properties frequently cause students confusion that persists into adulthood. To begin thinking about how to develop students' understanding of these properties, consider the problems in figure 4.29, guided by the question in Reflect 4.12.

Reflect 4.12

What ideas do the problems in figure 4.29 highlight?

1. I am thinking of two whole numbers that when multiplied give a product of 13. What can you tell me about the two whole numbers?
2. I am thinking of two whole numbers that when multiplied give a product of 0. What can you tell me about the two whole numbers?
3. Which is more, $6 \div 0$ or $0 \div 6$?

Fig. 4.29. Problems involving 1 and 0

Any of these problems would be appropriate to discuss with students in grades 3–5. All the problems are likely to lead to beneficial discussions in which students generalize and justify ideas about multiplying or dividing by 1 or 0.

In problem 1, since 13 has no other factors besides itself and 1, the only two factors that give a product of 13 are 1 and 13. The number 1 plays an important role in multiplication and division since 1 is a factor of any whole number—and note that 1 is a factor of 0. When introducing your students to the idea that "anything multiplied by 1 is that number," be sure to consider "switched" cases, such as 7×1 and 1×7. The first expression, which stands for "7 groups of 1," may confuse students, since many do not consider 1 object to be a group. They are likely to have an easier time with the second expression, which stands for "1 group of 7." Similarly, students should examine situations where a number is divided by 1 and consider how this differs from 1 divided by that number. Students should recognize that when they divide a number by 1, the result is that number.

In responding to problem 2, students should state that one or both of the numbers that are multiplied must be 0. Extending this situation to multiplication of more than two factors and a product of 0 means that at least one of these factors must be 0. This is an important idea that students will build on further when they study algebra. When introducing this idea to your students, be sure that they again consider the different meanings of "switched" cases, such as 4 × 0 and 0 × 4. Students should recognize that 4 × 0 means "4 groups with 0 in each group," whereas 0 × 4 means "0 groups of 4." Both result in 0, but the meanings are different. This investigation can lead to the conclusion that in any multiplication with 0, the product is 0.

Problem 3 broaches the subject of division that includes 0, a matter that can cause difficulty and confusion for many adults. Again, it is important that students think about what the expressions mean. Students can interpret the expression 0 ÷ 6 partitively, as meaning that they have nothing divided into 6 equal-sized parts, and they need to determine the size of each part—still 0. Another interpretation—a measurement interpretation—would involve thinking of having 0 objects and needing to determine how many groups of 6 can be made with 0 objects. Again, the result is 0—that is, the number of groups is 0. Thus, 0 ÷ 6 = 0.

Now consider the case of 6 ÷ 0. First, consider a measurement interpretation, beginning with 6 ÷ 2. How many jumps of 2 centimeters are needed to travel 6 centimeters? Clearly, 3 such jumps are needed. How about 6 ÷ 1? How many jumps of 1 centimeter are needed to travel 6 centimeters? Clearly, 6 jumps are required. What about 6 ÷ 0: How many jumps of 0 centimeters are needed to travel 6 centimeters? No matter how many jumps of 0 centimeters are taken, 6 centimeters will never be traveled.

In the same way, consider a partitive interpretation, beginning with 6 ÷ 2. If you partition a 6-centimeter piece of rope into 2 equal-length pieces, what will the length of each piece of rope be? Obviously, each piece of rope will be 3 centimeters. How about 6 ÷ 1? If you partition a 6-centimeter piece of rope into 1 equal-length piece, what will the length of each piece of rope be? Clearly, you will have 1 piece, and it will be 6 centimeters—the length of your original piece of rope. What about 6 ÷ 0? If, in the same way, you partition the 6-centimeter piece of rope into 0 equal-length pieces, what will be the length of each piece of rope—a partitive interpretation of 6 ÷ 0?

It is difficult to make sense of the partitive meaning, and such a discussion is not reasonable for an elementary classroom. However, the measurement division interpretation can serve to demonstrate that 6 ÷ 0 is undefined—that continued jumps of 0 centimeters would never result in jumping a distance of 6 centimeters. Such an interpretation can be used with elementary students to distinguish between 6 ÷ 0 and 0 ÷ 6.

Multiplication and division by 1 and 0 result in properties that are readily accessible for students in grades 3–5. However, these properties and ideas only emerge when students have a strong understanding of the meaning of multiplication and division. If students respond that 6 × 0 = 6, we should question the understanding that they are developing for the operation of multiplication.

Developing understanding of the right distributive property for division

Essential Understanding 2*b* highlights the fact that a form of the distributive property holds for division. This is worth noting, since many of the properties—commutative, associative, identity, and inverse—do not hold for division. Furthermore, this property forms the basis of the written algorithm for long division. Consider this fact as you respond to the questions in Reflect 4.13 about a fictitious student's claim, in the task shown in figure 4.30, that 721 ÷ 7 is the same as 700 ÷ 7 + 21 ÷ 7.

Reflect 4.13

Figure 4.30 shows the work of Mike, a fictitious student, who exhibits some understanding of the right distributive property for division in his claim that 721 ÷ 7 is the same as 700 ÷ 7 + 21 ÷ 7.

How is Mike's statement related to the long-division algorithm?

In what different ways would fourth- and fifth-grade students be likely to respond to Mike?

Mike said, "721 ÷ 7 is the same as 700 ÷ 7 + 21 ÷ 7."

Do you agree with Mike?

Fig. 4.30. A task designed to assess students' understanding of the right distributive property for division

We presented the task shown in figure 4.30 to a group of 134 fourth- and fifth-grade students. We found that students typically approached this problem by calculating both 721 ÷ 7 and 700 ÷ 7 + 21 ÷ 7. Whether or not they agreed with Mike depended on how they evaluated the second expression. Consider the work of Amaya, Rico, Troy, and Zoe, shown in figures 4.31, 4.32, 4.33, and 4.34, respectively.

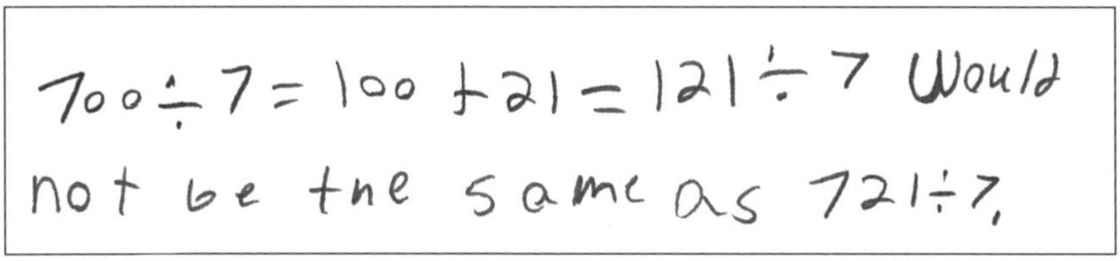

$$700 \div 7 = 100 + 21 = 121 \div 7 \text{ Would}$$

not be the same as $721 \div 7$.

Fig. 4.31. Amaya's (grade 5) response to Mike

You are right. I agree with you b/c
$700 + 21$ is the same thing as 721

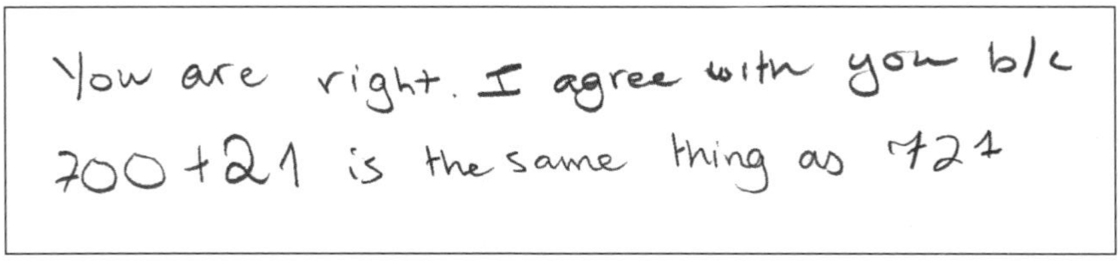

Fig. 4.32. Rico's (grade 5) response to Mike

I think it s $(700 + 21) \div 7 =$

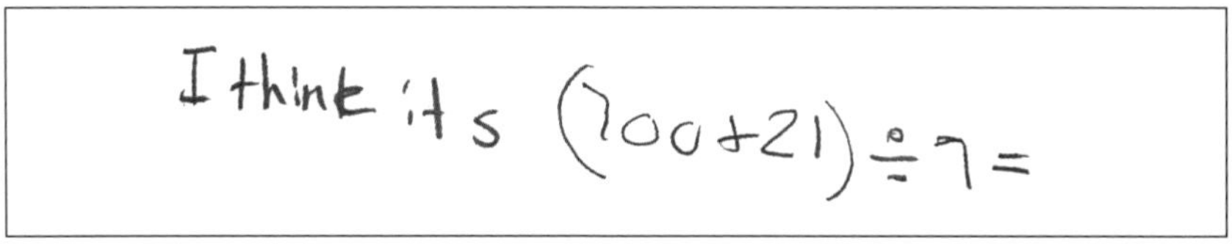

Fig. 4.33. Troy's (grade 5) response to Mike

No because it doesn't
have perinthases

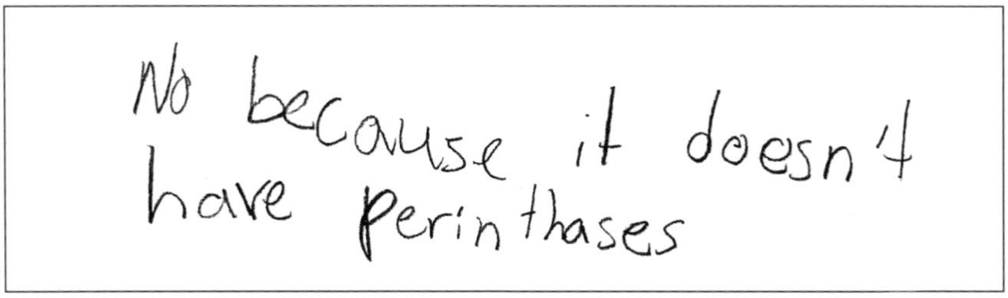

Fig. 4.34. Zoe's (grade 4) response to Mike

Only about 10 percent of the students (13 of 134) agreed with Mike. Some, like Rico (fig. 4.32), noted that 721 could be decomposed into 700 + 21. Many other students, however, disagreed with Mike. The majority of these students gave reasons similar to the explanation that Amaya provided (fig. 4.31), indicating that they calculated the second expression by working from left to right. Others, like Zoe (fig. 4.34), indicated that the two expressions would be equal if parentheses were inserted. Finally, several students agreed with Troy (fig. 4.33) and thought that Mike had written too much in the second expression.

On the basis of our sample of results, it appears that students did not follow the order of operations in evaluating the longer expression. It is possible that they had not yet studied this topic, but it is even more likely that they had had abundant experiences with binary operations (using two numbers) and fewer experiences with expressions using more than two numbers and multiple operations. This observation leads naturally to the order of operations, the focus of Essential Understanding 2c.

Developing understanding of order of operations

The two tasks shown in figure 4.35 appeared on the same assessment; they were sequenced in the order shown, but located on different pages. We asked fourth- and fifth-grade students to complete these two tasks. Examine the tasks and respond to the questions in Reflect 4.14.

Reflect 4.14

What information could you gain about your students' understanding of multiplication from their responses to the two tasks shown in figure 4.35?

In what ways might students' work on the two tasks provide you with different information about their understanding?

Task 1
What is the answer to $5 + 5 + 5 \times 0$?

Task 2
Patrick says that $6 + 5 \times 0$ is equal to 0 because $11 \times 0 = 0$.

Courtney says that $6 + 5 \times 0$ is equal to 6. She said,

"I took 5×0 first, which is 0. Then I took $6 + 0$, so the answer is 6."

Who is correct? Circle one: Patrick   Courtney

Fig. 4.35. Tasks designed to assess students' understanding of the order of operations

In response to the first question, 39 of the 40 students responded that $5 + 5 + 5 \times 0$ is equal to 0. Tyson, a fourth-grade student, gave a particularly clear explanation of how he obtained this value: "I figured it by doing $5 + 5 + 5 = 15$, then $15 \times 0 = 0$ in my head, and it equals 0 because the Zero Property of $\times$ says that anything $\times 0 = 0$." On the basis of students' work on this task, we were certain that they could multiply by zero, and it also seemed that they were not likely to be considering the priority of multiplication over addition in the order of operations.

Our results were slightly different on the second task, which presents two fictional students, Patrick and Courtney, who provide their reasoning on a similar computation: $6 + 5 \times 0$. Patrick's thinking illustrates the misconception evidenced by students in their work on the first task, while Courtney follows the order of operations correctly. In responding to this task, the majority of students (34 out of 40, or 85%) stated that Patrick was correct, often justifying their answer by saying that "anything multiplied by zero is zero." The remaining 6 students believed that Courtney was correct. One of those students was Monica, whose work is shown in figure 4.36. Monica's agreement with Courtney that $6 + 5 \times 0 = 6$ is at odds with her stated belief in the first task that $5 + 5 + 5 \times 0 = 0$. The questions in Reflect 4.15 probe the inconsistency in her responses to the two tasks.

Reflect 4.15

Presented with the first task shown in figure 4.35, Monica stated that $5 + 5 + 5 \times 0 = 0$. However, when she encountered the second task shown in the figure, she agreed with Courtney that $6 + 5 \times 0 = 6$.

How would you help Monica reconcile the conflicting views she has on these two tasks?

Many students justified their incorrect responses with a statement similar to "anything times zero is zero."

In what ways are these students correct? How could you help them refine their thinking so that they would respond correctly to these types of tasks?

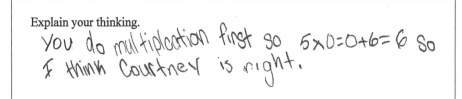

Explain your thinking.

You do multiplication first so 5x0=0+6=6 so I think Courtney is right.

Fig. 4.36. Monica's agreement with Courtney

Textbooks for elementary students may address the order of operations in a particular lesson. The typical lesson begins with a statement of the order of operations, and ends with several exercises providing students with practice in using the order correctly. The rationale, if any, given for this order is usually something like this: "Following this order guarantees that everyone, including students working with calculators, will get the same answer." Although this is a true statement, students may be left wondering why there wasn't a general agreement to proceed in some other order. Some may even notice that different calculators may evaluate the same expression in different ways.

Ameis (2011) suggests giving students opportunities to write expressions that involve more than one operation and are based on contextual situations. Figure 4.37 presents two such situations. Students can model these situations physically or pictorially, and in doing so, they can see why the operations of multiplication and division have priority over addition and subtraction. The task in figure 4.38 is designed to reverse this process by having students create a contextual situation for a given expression. Consider these tasks, guided by the questions in Reflect 4.16.

Reflect 4.16

What types of responses would you expect from students in grades 3–5 on the tasks in figures 4.37 and 4.38?

How would students' responses to the task in figure 4.38 help you assess their understanding of the order of operations?

- Debbie has 3 markers on her desk and 5 boxes with 8 markers in each box. Write an expression for the number of markers that she has altogether.

- Robbie, Daphne, and Angela shared a box of 24 crayons fairly. Robbie and Daphne also shared a box of 8 crayons fairly. Write an expression for the total number of crayons that Robbie has altogether.

Fig. 4.37. Situations that students could represent by symbolic expressions with multiple operations

Write a math problem about marbles that would be answered by using the expression

$5 \times 7 - 2 \times 4$.

Fig. 4.38. A task requiring students to create a situation that they could represent by a given expression with multiple operations

Ameis (2011) points out that the order of operations is not merely convention; some operations take precedence over others for mathematical reasons. At the same time, students may realize that in the expression $3 + 4 + 5 \times 0$, they can add the 3 and 4 before they multiply the 5 and 0, still yielding a correct result.

Summarizing Pedagogical Content Knowledge to Support Big Idea 2 through Essential Understandings 2a, 2b, and 2c

Teaching the mathematical ideas in this chapter requires specialized knowledge related to the four components presented in the Introduction: learners, curriculum, instructional strategies, and assessment. The four sections that follow summarize some examples of these specialized knowledge bases in relation to Big Idea 2, which in this chapter is treated primarily through Essential Understandings 2a, 2b, and 2c. Although we separate the four knowledge bases to highlight their importance, we also recognize that they are connected and support one another.

Knowledge of learners

This chapter has discussed samples of student work that illustrate several common misconceptions that research (for example, Greer 1994) has highlighted. For instance, students often state that multiplication of two numbers always yields a larger number or that the dividend must be greater than the divisor. They frequently think that commutativity always applies in division situations and that division by 0 is possible. The tasks presented in this chapter are designed to bring these misconceptions to the surface so that teachers can identify and challenge them in productive ways.

Knowledge of curriculum

In addition to carefully selecting tasks that do not perpetuate misconceptions, you may need to enhance tasks that you find in your curricular materials. As the chapter has discussed, textbooks often give undue emphasis to symbolic representations when introducing properties related to multiplication and division. Therefore, it is important to create related contextual situations that will help students build essential understandings. Kabiri and Smith (2003, p. 187) explain:

> Textbook problems do not always lend themselves to multiple solutions, or solution strategies. However, many problems can be made more open-ended and accessible to a wide variety of student ability levels with minimum effort.

Consider your curricular goals and your students as you enhance mathematical tasks in curricular materials. Ask yourself, for example, whether the tasks that you are using—

- have a design that facilitates classroom analysis and debate of student misconceptions;

- encourage multiple solutions or solution strategies;

- have language that is misleading or confusing.

Ideas for enhancing curriculum materials are available in Chval and Davis (2008), Chval and colleagues (2009), Chval and Chavez (2011/2012), and Kabiri and Smith (2003).

Knowledge of instructional strategies

Teachers can draw on many and varied instructional strategies in helping their students interpret multiplicative situations and work with models and representations of them. We highlight two examples here. First, as you facilitate discussions about your students' representations, emphasize the meaning that lies behind the representations' various components. Connecting symbolic and pictorial representations, as illustrated in figure 4.16, can help students develop their understanding of the meaning of properties of multiplication. Second, provide opportunities for your students to use physical tools to model situations. For example, for problems involving division by 0, create a number line on the floor by using painter's tape, and ask students to make the physical jumps to explore situations that you propose to them.

Knowledge of assessment

Writing can help students develop their understanding of mathematical concepts (Shepard 1993) and can also be highly useful in giving teachers access to students' understanding of mathematics (Pugalee 1997; Silver, Kilpatrick, and Schlesinger 1990). Furthermore, mathematical writing encompasses different genres. For instance, Marks and Mousley (1990) describe the following genres:

- Procedure—writing that tells how something is done

- Description—writing that tells what a particular thing is like

- Report—writing that tells what an entire class of things is like

- Explanation—writing that tells the reason why a judgment has been made

- Exposition—writing that presents arguments about why a thesis has been produced

This chapter has discussed some tasks that asked students to explain their thinking. It is important to help students develop writing competencies that are analogous to those that mathematicians use. Therefore, you need assessment tasks that require your students not only to describe how they solved mathematics problems, but also to provide mathematical arguments, justify their thinking, and generalize beyond a small number of cases. Useful assessment questions to ask your students to assess their thinking include the following:

- "Why does this procedure work?"

- "Will this strategy always work?"

- "How do you know?"

Responding to such questions will help students develop skill in using different genres of mathematics writing while at the same time providing you with opportunities to assess their level of skill and understanding.

Conclusion

In summary, the properties of multiplication and division are powerful tools for solving problems. Although we have discussed the properties separately, they are often used in conjunction with one another in mental arithmetic. For example, to find 40×7, many adults will multiply 4×7, and then multiply that result by 10. This correct strategy uses both the associative and commutative properties. Furthermore, these properties provide the rationale for various algorithms for multiplication and division—the subject of the next chapter.

into
practice

Chapter 5
Algorithms for Multiplication and Division

Essential Understanding 2*d*
Properties of operations on whole numbers justify written and mental computational algorithms, standard and invented.

Developing efficient algorithms for multiplication and division requires considerable attention in grades 3–5. As noted by Bass (2003), algorithms are helpful tools that are an important part of mathematics. Efficient written and mental algorithms rely on properties of operations, an idea presented as Essential Understanding 2*d* in *Developing Essential Understanding of Multiplication and Division for Teaching Mathematics in Grades 3–5* (Otto et al. 2011). Algorithms allow students to determine a result in an efficient manner when modeling a situation directly would become a tedious chore (Lampert 1986).

Thus, it is critical that you allow your students the time necessary to develop and invent strategies for multi-digit multiplication and division and that you move your students toward efficient strategies. However, you must also recognize that most calculations that they will do as adults will involve the use of everyday, readily available technologies. In addition, you must be cautious about moving them too quickly to the use of sophisticated algorithms. Steffe (1994) synthesized research on students' use of multiplication algorithms, noting that students often refuse to use the procedures that are taught in school and continue to use their own intuitive and frequently cumbersome strategies. This chapter focuses on rethinking and restructuring ways of introducing students to algorithms in elementary school.

Working toward Essential Understanding 2d

As noted by Van de Walle (2007), "As much time as necessary should be devoted to the [underlying ideas] of the algorithm with the recording or written part coming later" (p. 234). Some would say that rushing students into using an algorithm that they do not understand is no better than simply handing them a calculator, except that the calculator will actually produce correct results! To begin to examine the role that algorithms play in multiplication and division in grades 3–5 classrooms, consider the question in Reflect 5.1.

Reflect 5.1

What are important characteristics of the multiplication and division algorithms that you teach in your classroom?

Bass (2003, pp. 323–24) identified the following qualities that teachers should consider when determining the usefulness of algorithms:

(a) *Accuracy*. The algorithm always produces a correct answer.

(b) *Generality*. The algorithm applies to all instances of the problem, or class.

(c) *Efficiency*. The cost (in time, effort, difficulty, or resources) of executing the algorithm is reasonably low compared with the input size of the problem.

(d) *Ease of accurate use* (versus error proneness). The algorithm can be used reasonably easily and does not have a high frequency of error in use.

(e) *Transparency* (versus opacity). What the steps of the algorithm mean mathematically and why they advance us toward the problem solution are clearly visible.

The first three qualities are those that most adults would be likely to identify. Certainly, algorithms should always produce correct results and use as little time or effort as possible. However, the last two qualities on Bass's list are particularly important to teachers: students need algorithms that they can learn and use without introducing errors that can be difficult to overcome. Further, they need to have a clear understanding of the underlying mathematical ideas that support an algorithm.

In addition to the five qualities identified by Bass, a sixth quality is important to teachers: an algorithm should connect with and facilitate the students' future learning of mathematics. Algorithms that connect with procedures that students will learn later and use in algebra can facilitate the transfer of these ideas and reduce instructional time needed for them subsequently in the school mathematics curriculum.

Building understanding of algorithms for multiplication

In considering how to develop students' understanding of an efficient and effective algorithm for multiplication, recognizing some of the pitfalls that researchers have noted can be helpful. Van de Walle (2007) observed that "the traditional multiplication algorithm is probably the most difficult of the four algorithms if students have not had plenty of opportunities to explore their own strategies" (p. 234). Because teachers, along with many other adults, have a much deeper understanding of algorithms than students, they often fail to recognize the difficulties that students experience. Consider the four fourth-grade students' use of multiplication algorithms for 49 × 40 shown in figures 5.1–5.4. Use the questions in Reflect 5.2 to guide your inspection of the students' work.

Reflect 5.2

Figures 5.1–5.4 show the algorithms that four fourth-grade students used to compute 49 × 40.

What do these students appear to understand about multiplication?

Would these algorithms work for multiplying other numbers as well, such as 24 × 17?

In what ways do these algorithms make some sense, even if they are incorrect?

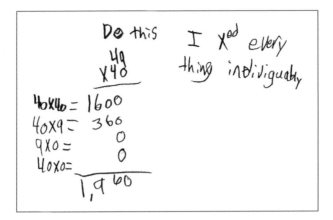

Fig. 5.1. Brady's (grade 4) use of an algorithm for 49 × 40

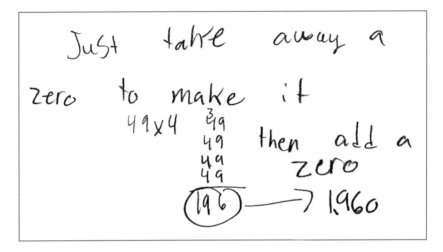

Fig. 5.2. Susanna's (grade 4) use of an algorithm for 49 × 40

Fig. 5.3. Micah's (grade 4) use of an algorithm for 49 × 40

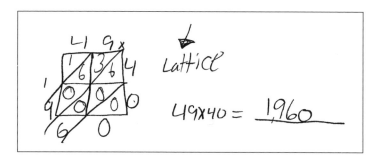

Fig. 5.4. Veronica's (grade 4) use of an algorithm for 49 × 40

We asked 61 fourth-grade students to find the product of 49 and 40. Although these students had not yet received instruction on multi-digit multiplication algorithms, about half of them (31 out of 61) chose to invent an algorithm. Most of these algorithms resulted in an incorrect product (18 out of 31, or 58 percent). Micah (fig. 5.3) demonstrated the most frequently used incorrect algorithm, which involved "stacking" the factors and multiplying down the columns. This method mirrors algorithms used for multi-digit addition and subtraction, and its similarity to them may explain why our students frequently used it. In fact, one of the challenges of teaching multiplication algorithms to students is helping them understand that multiplying down the columns does *not* generate the correct result. However, students often try to apply previously learned ideas to new situations, as they did in this case.

Brady, Susanna, and Veronica all obtained the correct product with their algorithms, which were very different from one another. To compute 49 × 40, Susanna (fig. 5.2) considered the related expression 49 × 4, used repeated addition of four 49s to obtain 196, and then "add[ed] a zero" to arrive at a product of 1,960. Susanna's teacher might have asked her whether this method of removing the ones digit from a factor and attaching it to the end of the product obtained without that digit would work in other cases. For example, could Susanna find 24 × 17 by using 24 × 1 = 24, and then attaching a 7 to the end of 24 to get 247? Although language related to addition and subtraction is commonly heard in classrooms in this context, Susanna was not "taking away" or "adding" 0 to modify the problem or the product; instead, she was dividing or multiplying by 10. It is possible that Susanna understood that her work relied on the associative property: 49 × 40 = 49 × (4 × 10) = (49 × 4) × 10. However, whether she did would not be clear without further probing. It is essential to help your students see the connections between properties and algorithms and to model the correct use of language for them.

Current textbooks approach multiplication algorithms in a variety of ways. Two common algorithms in textbooks are the partial products algorithm and the lattice algorithm. For more information on these algorithms, see Randolph and Sherman (2001) and Otto and colleagues (2011). Both algorithms ensure that each digit of the first factor is multiplied by each digit of the second factor. Brady (fig. 5.1) demonstrated the partial products algorithm and actually included the factors used in determining the four partial products. Of the work shown in figures 5.1–5.4, Brady's procedure exhibits the deepest understanding of multiplication. Veronica (fig. 5.4) used the lattice algorithm correctly, but her work does not reveal whether she understands why the algorithm works.

It is important to note that an ability to reason multiplicatively, knowledge of multiplication algorithms, and skill in using algorithms successfully are not the same; in fact, it is possible to have one without the others. The goal is to ensure that students have all three. Later, students will multiply and divide numbers that extend beyond the whole numbers, including rational numbers expressed as decimals with many digits to the right of the decimal point. If students do not understand the underlying place-value ideas used in their algorithms, they are likely to make errors when they attempt to extend earlier algorithms to new situations.

We showed the task that appears in figure 5.5 to the same group of fourth-grade students and a group of fifth-grade students at the same school. The task presents a fictitious student, Marsha, in the process of using the traditional algorithm to multiply two two-digit numbers. Examine the task in figure 5.5 and respond to the questions in Reflect 5.3.

Reflect 5.3

The Traditional Multiplication Algorithm task in figure 5.5 shows students a snapshot of work in progress by a fictitious student, Marsha, as she computes 37×54. The students answer questions about key aspects of Marsha's work.

How would you respond to the questions posed to students about the meaning of the 2 and the 0 that Marsha has written as she works with the algorithm?

What various responses would you anticipate from your students *before* they learned the traditional algorithm for multi-digit multiplication?

What various responses would you expect from your students *after* they learned the traditional algorithm for multi-digit multiplication?

> Marsha was multiplying 37 × 54. Marsha wrote the following:
>
> 2
> 37
> × 54
> ———
> 148
> 0
>
> **What does the little "2" written above the 3 in 37 mean?**
>
> **Why did Marsha put a "0" below 148?**

Fig. 5.5. Traditional Multiplication Algorithm task

Because these fourth-grade students had not yet learned an algorithm for multi-digit multiplication, it is not surprising that 39 percent (24 out of 61) stated that they did not know the meaning of the 2, and 62 percent (38 out of 61) said they did not know why Marsha wrote a 0 under the 8 in 148. About half of the fourth-grade students reasoned that the 2 represented 20 in 28, which was obtained by multiplying 7 and 4. The fifth-grade students had previously learned the traditional algorithm and had little difficulty with the meaning of the 2. The most common responses about the 0 were similar to those shown in figure 5.6.

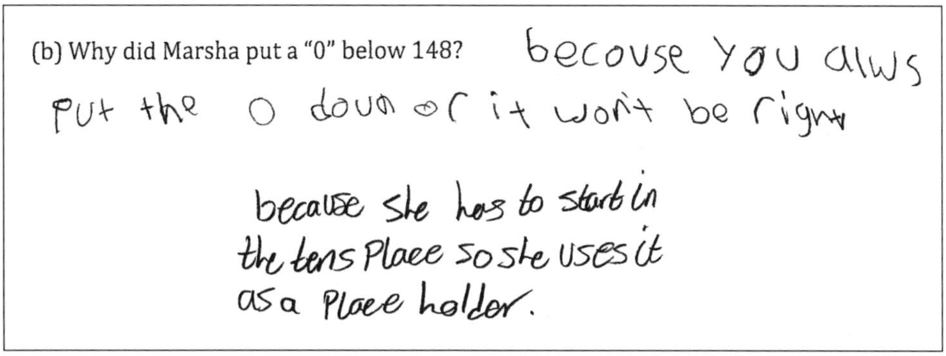

> (b) Why did Marsha put a "0" below 148? becouse you alws put the 0 doun or it wont be right
>
> because she has to start in the tens Place so she uses it as a Place holder.

Fig. 5.6. Common responses from fifth-grade students about the meaning of the 0 in the Traditional Multiplication Algorithm task

Almost 80 percent (46 out of 58) of the fifth-grade students stated that the 0 is necessary for the answer to be correct or that the 0 is a placeholder. On the basis of what these students wrote, it is difficult to determine the depth of their understanding of the meaning or purpose of the 0. Although the term *placeholder* is not typically found in today's textbooks, it does live on in the vernacular of teaching and appears on a number of teaching websites. Along with the terms *carrying* and *borrowing*, *placeholder* should be eliminated from the classroom, with teachers instead focusing students' attention on the actual meaning and value of 0. In the traditional algorithm for multiplication, the 0 does more than simply hold a place; it shows that the value of the ones place is 0 in multiplications by tens. The fifth-grade student whose work is shown in figure 5.7 demonstrates an emerging understanding of the meaning and purpose of 0 in the algorithm.

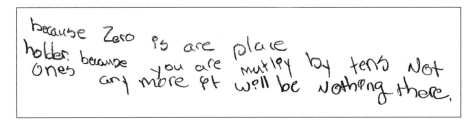

Fig. 5.7. A fifth-grade student's emerging understanding of the meaning of the 0 in the Traditional Multiplication Algorithm task

Nearly ninety years ago, Buswell (1926) stated that students who knew their basic multiplication facts demonstrated various common errors in working with the traditional multiplication algorithm. Decades later, Lampert (1986) emphasized that students often make these common errors not because they lack an understanding of place value but because they do not realize that they should consider it in using the algorithm. It is useful to be aware of and avoid the difficulties that arise from introducing students to algorithms too soon, without the underlying support of the mathematical ideas.

Van de Walle (2007), Lampert (1986), and others suggest introducing students to multiplication algorithms through contextual situations that promote the use of an array and involve a two-digit multiplicative unit and a single-digit multiplier—for example, 6 × 38. Reflect 5.4 offers one such contextualized task as an entry point to a closer examination of this approach.

Reflect 5.4

Consider the following contextualized task:

In the school gymnasium, there are 6 rows of chairs and 38 chairs in each row. How many chairs are in the gymnasium?

What strategies and representations would you expect students to use in approaching this task?

One possible strategy for use with this contextualized task would be direct modeling, using physical items or a drawing, as shown in figure 5.8. Although students could find the total number of chairs by counting, considering each row as 30 chairs and 8 chairs might be more efficient. This way of thinking is represented in figure 5.9 by shading.

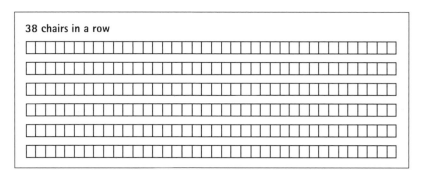

Fig. 5.8. A rectangular array of 6 rows of chairs with 38 chairs in each row

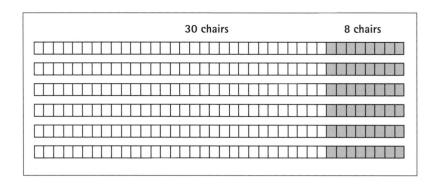

Fig. 5.9. A rectangular array of 6 rows of chairs with 38 chairs in each row, but with the last 8 chairs in each row shaded

The array in figure 5.9 uses shading to highlight 6 groups of 30 chairs and 6 groups of 8 chairs. The 6 groups of 30 can also be seen as 6 groups of 3 tens, or 18 tens, or 180 chairs. The 6 groups of 8 chairs contain 48 chairs, so the total number of chairs is the sum of these parts: 180 + 48 = 228. Figure 5.10 connects this diagram and description to the partial products algorithm (shown in two versions) and the traditional algorithm for multiplication. Examine these algorithms and respond to the questions in Reflect 5.5.

Reflect 5.5

Figure 5.10 shows the traditional algorithm and two ways of using the partial products algorithm to find 6 × 38, the multiplication representing the chair context presented in Reflect 5.4.

How do the algorithms in figure 5.10 connect with the direct modeling shown in figure 5.9?

How do these algorithms compare with respect to Bass's (2003) qualities of accuracy, generality, efficiency, ease of accurate use, and transparency?

Two versions of the partial products algorithm are shown. In what situations would you prefer to use one rather than the other?

Partial products algorithm		Partial products algorithm		Traditional algorithm
38		38		$\overset{4}{38}$
× 6		× 6		× 6
180	30 × 6	48	8 × 6	228
+ 48	8 × 6	+180	30 × 6	
228		228		

Fig. 5.10. Algorithms used to calculate 6 × 38

The first two methods are called *partial products algorithms* because they display the underlying, intermediate calculations needed to compute 6 × 38. Both the partial products and the traditional algorithms are accurate and can be generalized for multiplying any two whole numbers. They differ, however, in efficiency, ease of accurate use, and transparency. In many instances, the traditional algorithm uses less space on paper but may require more time and mental effort to complete. The partial products algorithm is more transparent than the traditional algorithm in that the partial products are listed vertically below the factors. Note that the traditional algorithm combines the steps in the partial products algorithm, using the little 4 above the tens column to indicate the 4 tens that are "regrouped" above the 3 in 38.

A number of challenges emerge when teaching the traditional algorithm for multiplication. Students may become confused about the meaning of the 4 and what they should do with it (should they add it to or multiply it with the product of 3 and 6?). Van de Walle (2007) asserted that "there is absolutely no practical reason why students cannot be allowed to record partial products and avoid the errors related to the carried digit" (p. 235). Furthermore, the traditional algorithm does not encourage students to consider place value, whereas the partial products algorithm does. Finally, the partial products algorithm connects with mathematics that the students will learn in later grades, such as the multiplication of polynomials.

The two partial products algorithms shown in figure 5.10 differ in the order in which they list the partial products. Because of the commutative property of addition and because both of these methods preserve place value, they always result in the same answer. The first, which is sometimes referred to as a left-to-right method, is generally easier to use in mental computation and can help students generate reasonable estimates for their results because this method considers larger items first—here, the 30 before the 8. In addition, the left-to-right method mirrors reading from left to right and therefore gives better support to struggling learners. The second partial products algorithm is sometimes used as a bridge to help students make the transition to the traditional algorithm, which is a right-to-left method, calling for multiplying the ones digits first.

In a summary of research on teaching and learning multi-digit multiplication and division, Fuson (2003a, 2003b) recommends that teachers begin by using area models that have been subdivided as arrays and using these models to highlight the results of multiplying by 1, 10, and 100. She suggests gradually helping students move from these early models using arrays to a more abbreviated model once they understand the meaning of the various components of the multiplication algorithm. Figure 5.11 shows the first-stage model, which Fuson calls an "array-size model," as well as the second-stage, "abbreviated" model.

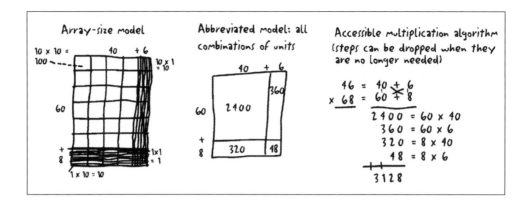

Fig. 5.11. Multiplication algorithms from Fuson (2003b, p. 303)

Students can create the first-stage model as they try to determine a way to calculate the subcomponents of the array. Teachers can then introduce the abbreviated model (if a student does not) and connect it with the accessible partial-products algorithm, shown on the right in figure 5.11.

Building understanding of algorithms for division

As in teaching algorithms for multiplication, Fuson (2003b, p. 303) recommends beginning with arrays in teaching algorithms for division. For example, figure 5.12 shows four algorithms for dividing 3,129 by 46, including an abbreviated model for the division.

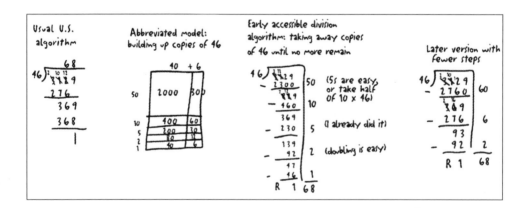

Fig. 5.12. Division algorithms from Fuson (2003b, p. 303)

The usual U.S. algorithm (also called the *long-division algorithm*) is very different from the algorithms for the other operations in two ways: (a) the process works from left to right, and (b) students write their answers and work in a number of different positions. The algorithm is also not mathematically transparent—for example, the connection with place value is not clear—and the algorithm is sometimes error prone. According to Fuson (2003b, pp. 303), this algorithm requires students to "determine exactly the maximum copies of the divisor that they can take from the dividend." She elaborates:

> This feature is a source of anxiety because students often have difficulty estimating exactly how many will fit. ... [T]he current algorithm creates no sense of the size of the answers that students are writing; in fact, they are always multiplying by single digits. In the example in [fig. 5.12], they just write a 6 above the line; they have no sense of 60 because they are literally only multiplying 46 by 6. Thus, students have difficulty gaining experience with estimating the correct order of magnitude of answers in division when they are using the current U.S. algorithm. (p. 304)

To help students develop an understanding of division algorithms, Fuson recommends earlier work with models that are based on arrays and incorporate ideas from measurement division, like the abbreviated model shown in figure 5.12. Notice that the array-based models for multiplication and division are similar, demonstrating an important connection between these two operations. In the example shown in figure 5.12, students build the model with a side of length 46, split into two parts of size 40 and 6. Students then build the other side of the model, using multiplication to determine the partial products. They continue this process until they can add no more copies of the divisor or multiplicative unit (46 in this case) without exceeding the dividend (3,129). The difference of the dividend and partial products is the remainder, allowing the students to see why the remainder must be less than the divisor but greater than or equal to 0.

The two algorithms on the right in figure 5.12 are sometimes called *scaffold algorithms*, and are included in some textbooks. Compare the scaffold algorithms with the abbreviated model in figure 5.12 and respond to the questions in Reflect 5.6.

Reflect 5.6

Where is the quotient found in the abbreviated model in figure 5.12?

The first two partial products in the early accessible division algorithm are 2,300 and 460. Where are these found in the abbreviated model? Where are they located in the second scaffold algorithm (the later version, with fewer steps)?

One important feature of the scaffold algorithm is that students may choose differ- ent partial products while building up copies of the divisor, but they are still able to determine the quotient accurately. Examine the work shown in figure 5.13 from two fifth-grade students, and then respond to the question in Reflect 5.7.

Reflect 5.7

Amy and Boyd, the two fifth graders whose work is shown in figure 5.13, arrive at different answers for 200 ÷ 6.

How would you help the student who did not obtain the correct quotient?

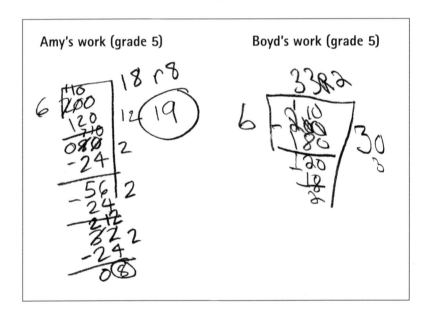

Fig. 5.13. Two fifth graders' use of the scaffold algorithm for 200 ÷ 6

Amy subtracted 120 initially but seems to have been confused about how many 6s she needed to make 120, writing 12. She then successively subtracted three 24s, ei- ther thinking that two 6s make 24 or else using the 12 that she wrote at the begin- ning (two 12s make 24). Boyd recognized that he could subtract thirty 6s (a total of 180) from 200 first and then three more 6s (a total of 18), leaving a remainder of 2, so he wrote the quotient as 33 R2. The scaffold division algorithm allows flexibility in the process and provides an opportunity for teachers to encourage students to take fewer and fewer steps as they become more efficient in their decisions about which amounts to subtract.

Summarizing Pedagogical Content Knowledge to Support Essential Understanding 2*d*

Teaching the mathematical ideas in this chapter requires specialized knowledge related to the four components presented in the Introduction: learners, curriculum, instructional strategies, and assessment. We summarize some examples of these specialized knowledge bases in the following four sections. Although we separate them to highlight their importance, we also recognize that they are connected and support one another.

Knowledge of learners

It is critical to allow students the time necessary to develop initial strategies for multi-digit multiplication and division and move students toward efficient strategies. Students often struggle with developing essential understandings related to multiplication and division algorithms. As illustrated by Micah's work in figure 5.3, students may misapply steps that they have learned for addition and subtraction algorithms when they begin working with multiplication and division. Moreover, they may know the correct steps without having developed an understanding of their meaning. These are important issues to anticipate, address, and assess.

Knowledge of curriculum

Current textbooks approach multiplication algorithms in a variety of ways. For example, two multiplication algorithms that appear in textbooks are the partial products algorithm and the lattice algorithm. Which algorithms do your curricular materials introduce? What models and representations do they use to help students build their understanding of the meaning of multiplication and division algorithms? Do students have opportunities to create their own representations or discuss the meaning of the different components of algorithms? Are students introduced to the important qualities of algorithms, such as efficiency and transparency (Bass 2003)? Fuson (2003a, 2003b) suggests a sequence for introducing algorithms to help students develop essential understandings and proficiencies. Do your materials have a curricular sequence that makes sense and provides sufficient time for students to develop these ideas?

Knowledge of instructional strategies

Teachers can draw on many and varied instructional strategies in helping their students develop understanding of multiplication and division algorithms. A few examples are highlighted here.

It is important that you help your students see the connections between properties and algorithms and that you model the correct use of language for them. For example, you might use a representation such as that of the chairs context in figure 5.9 to help your students make the connection from the representation to the partial-products algorithm, displayed in figure 5.10, and the properties discussed in Chapter 4.

You should also be careful not to use language such as Susanna used in her work, displayed in figure 5.2. In using an algorithm for multiplication, Susanna talks about "add[ing] a zero" and "tak[ing] away a 0," phrases that can cause confusion of multiplication with addition and subtraction, especially for English language learners.

Finally, Van de Walle (2007), Lampert (1986), and others suggest introducing students to algorithms through contextual situations that promote the use of arrays. Reflect 5.4 presents one such task, set in the context of rows of chairs in the gymnasium.

Knowledge of assessment

It is important to give students tasks that provide you with opportunities to assess whether they have developed essential understandings related to the different components of multiplication and division algorithms. Such tasks might pose questions like those in the Traditional Multiplication Algorithm Task in figure 5.5, which calls for students to explain the meaning of particular digits written by a fictitious student in the process of using the multiplication algorithm. Or, as in Reflect 5.6, assessment tasks might pose questions that ask where the quotient and partial products are located in the abbreviated model for division or the different versions of the division algorithm. Tasks that require students to analyze student representations of algorithms or explain the meaning of specific steps that are displayed can play different roles and serve different purposes from items that require students to use algorithms only to make accurate and efficient computations.

Conclusion

The recommendations from Fuson (2003a, 2003b) and others demonstrate the value of introducing students to algorithms for multiplication and division with contextual situations and diagrams that model them. In particular, arrays are helpful in illustrating products for the teaching of algorithms. You can then connect these models with written procedures and move toward accessible algorithms, such as the partial products algorithm. For many students, the accessible algorithms are sufficient for the calculations that they will make. However, you should also introduce the traditional algorithms and make the steps in these procedures transparent to your students. As you introduce students to these algorithms, keep in mind the qualities of useful algorithms that Bass (2003) identifies. The goal is to help your students understand the supporting mathematical ideas for the algorithms, aiding their long-term retention and providing them with the foundational understanding necessary for their future success.

The next chapter highlights the alignment of the essential understandings discussed in the preceding chapters with mathematical concepts and topics that students encounter at lower and higher levels of learning.

Chapter 6
Looking Back and Looking Ahead with Multiplication and Division

This chapter highlights how the essential understandings discussed in Chapters 1–5 align with ideas that students develop before and after grades 3–5. The chapter begins with a discussion of foundational understandings that students are expected to build in kindergarten through grade 2. When you encounter students who have gaps in their knowledge in grades 3–5, you may need to assess their understanding of the ideas that this first section highlights. The second section discusses how the essential understandings presented in Chapters 1–5 connect with mathematics that students learn beyond fifth grade. This discussion demonstrates how important it is for students in grades 3–5 to develop a deep understanding of the essential concepts that serve as a foundation for subsequent learning.

Supporting Knowledge in K–Grade 2 for Multiplication and Division in Grades 3–5

Teachers in kindergarten through grade 2 can support the development of students' essential understandings of multiplication and division in grades 3–5. For example, in these grades it is important to help students establish a foundational understanding of the meaning of a unit, develop multiplicative thinking, and learn to use various representations in ways that they can build on in grades 3–5. It is useful to discuss these aspects of foundational understanding in greater detail.

Focusing students on the unit

As students learn to count and measure, it is important to emphasize the unit that they are using and help them recognize that the count or measure changes, depending on the unit that is established. For example, students should recognize that if they hold up their hands, they can count the number of hands or the number of fingers, but the count will vary, depending on the unit (see fig. 6.1). Furthermore, when students play games such as concentration, they create pairs of cards that are "matches." When all the matches are made and the game ends, they determine the winner by counting either the number of pairs or the number of cards (see fig. 6.2). For their determination to be correct, they need to be consistent in their use of the unit that they establish.

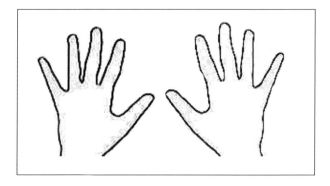

Fig. 6.1. Diagram of 2 hands or 10 fingers, depending on the unit chosen

Fig. 6.2. A model of 2 pairs or 4 cards in the game concentration

Clarifying the multiplicative unit that is counted is a key component of the skip counting process. Often the multiplicative unit is assumed, though it is important to emphasize that typically when we count, we determine a quantity on the basis of some identified unit. Counting situations arise in a variety of contexts, and discussions can occur quite naturally about when one count or measure is more than, less than, or the same as another, providing opportunities to introduce students to the use of the language needed for comparing counts in grades 3–5. Changing the size of the multiplicative unit changes the counting sequence (e.g., skip counting by fives: 5, 10, 15, 20; and skip counting by threes: 3, 6, 9, 12, 15). Students should consider the relationship between the size of the multiplicative unit and the count or measure.

As students engage in situations that involve multiplication and place value, they must look at units in a new way, often referred to as *unitizing*. Unitizing means viewing objects or quantities in different-sized chunks (Lamon 1999). For example, unitizing would involve viewing the diagram in figure 6.3 as 30 blocks and as 3 groups of 10 blocks. Similarly, students would recognize 5 groups of 4 objects also as 20 objects. Depending on the unit or group, they could count the number of groups or the total number of objects.

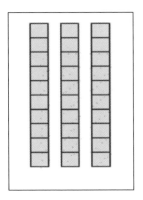

Fig. 6.3. Three groups of 10 blocks or 30 blocks

Measurement often provides a natural connection and need for establishing a unit. Depending on what you are measuring, you need to select an appropriate unit, and then you reason on the basis of that unit. For example, you have various units to choose from when measuring length, including inches, feet, miles, meters, kilometers, and light-years. You could also choose to measure the area, temperature, brightness, or humidity of your classroom. As you consider what attribute to measure, you need to select a unit that is appropriate for the attribute. Measuring the length of the classroom in shoe-lengths will result in a different count from measuring the length of the classroom with units the length of paper clips. Discussions about the importance of the selection of a unit can help students understand the relationship between the unit and the count, providing a bridge to multiplicative reasoning.

Multiplicative thinking

Students in grades K–2 often reason in an additive manner. For example, when given 5 marbles, they consider how many more marbles they need to have 12 marbles. However, students in these early grades can also engage in multiplicative thinking, particularly in measurement situations in which the unit changes, but also in situations that involve equal groups of objects.

The Common Core State Standards for Mathematics (CCSSM; National Governors Association Center for Best Practices and Council of Chief State School Officers [NGA Center and CCSSO] 2010) emphasize the importance of helping students in grade 2 lay the groundwork for multiplicative thinking. Figure 6.4 shows CCSSM's expectations that second graders will work with groups, pairing objects or counting them by 2s, and will work with rectangular arrays.

Common Core State Standards: Grade 2

Work with equal groups of objects to gain foundations for multiplication.

3. Determine whether a group of objects (up to 20) has an odd or even number of members, e.g., by pairing objects or counting them by 2s; write an equation to express an even number as a sum of two equal addends.

4. Use addition to find the total number of objects arranged in rectangular arrays with up to 5 rows and up to 5 columns; write an equation to express the total as a sum of equal addends.

Fig. 6.4. Operations and Algebraic Thinking, CCSSM 2.0A.3 and 4
(NGA Center and CCSSO 2010, p. 19)

It is important that we distinguish between multiplicative thinking and the process of multiplication. Confrey and Harel (1994) explain that reasoning mathematically about situations means reasoning about *things and relationships*. Students need to reason about things and relationships to develop and deepen their use of multiplicative thinking. Jacob and Willis (2001) emphasize that children must first come to recognize that multiplicative situations involve three aspects: groups of equal size, a number of groups, and a total amount. Figure 6.5 provides a representation of these three aspects that students must recognize: (a) groups that are equal in size (in this case, groups containing 4 circles), (b) the number of groups (in this case, 3 groups), and (c) the total number in all the groups (in this case, 12 circles). Furthermore, when children can construct and coordinate these three factors in problem situations, they are thinking multiplicatively. The goal is for students to recognize when to and when not to reason multiplicatively, given a particular situation.

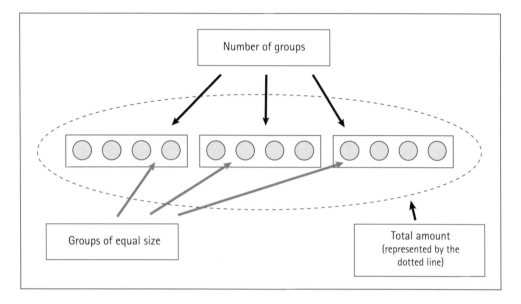

Fig. 6.5. The three aspects of a multiplicative situation

Jacob and Willis (2001, p. 306) also argue that teachers need to recognize the difference between additive and multiplicative thinking:

> Multiplication is more than repeated addition, however, and its learning more complicated. While repeated addition may be an appropriate beginning, to maintain that interpretation of multiplication is ultimately disabling because it does not provide children with important multiplicative structures.

These words of caution about interpreting multiplication as repeated addition emphasize an important idea about the meaning established for multiplication. Students should recognize that 3×5 means 3 equal-sized groups of 5, rather than the often-introduced meaning of 3×5 as 5 added 3 times (that is, $5 + 5 + 5$). As students move toward developing an understanding of the meaning of fractions and of multiplication of fractions, they will discover that finding $^2/_3$ of a group of $^1/_2$ (that is, $^2/_3 \times ^1/_2$) makes sense, but that adding $^1/_2$ two-thirds times does not. Viewing multiplication as a relationship between the number of groups and the number in each group allows for a smooth transition to the meaning of fraction multiplication.

Consider the following problem:

> You have 20 marbles. You would like to put the marbles into 4 bags, with the same number of marbles in each bag. How many marbles should you put into each bag?

This problem requires students to partition 20 marbles into 4 equal-sized groups. Such problems provide young students with a foundation for the partitioning that they will encounter later when they are introduced to fractions. Battista (2012) stresses the point: "Before students can understand fractions, they must understand partitioning. To partition a whole is to divide it into equal portions, like dividing a pizza equally among four people" (p. 1).

The difficulty of sharing problems can vary, depending on the numbers involved, the model in use (for example, set or area), and the presence of physical manipulatives (Van de Walle 2007). According to Van de Walle, children's initial strategies for sharing typically involve halving, so problems involving 2, 4, or 8 sharers are a good place to start. For example, students could demonstrate how to share 10 cookies evenly among 4 people to consider situations that involve parts of wholes.

Contexts that introduce students to the meaning of "doubling" and "halving" and extending to situations that involve increasing an amount by 3 times (tripling) or 4 times (quadrupling) encourage multiplicative reasoning. For example, the following situation promotes multiplicative reasoning:

> Sarah has 8 toy cars. Martha has 2 times as many toy cars as Sarah. How many toy cars does Martha have?

Such situations promote *iterating* the group—that is, *creating duplicates* of a unit in a process similar to what students must do when they create $2/8$. The toy car problem refers to two groups of 8 objects, whereas a situation involving $2/8$ refers to two groups of $1/8$.

Both iterating and partitioning are critical processes that students use in developing an understanding of multiplicative reasoning (Siebert and Gaskin 2006). Tasks that encourage students to partition groups of objects into equal-sized groups or to iterate groups of objects promote the use of multiplicative reasoning. Students can iterate and partition continuous objects as well as discrete ones. Consider the tasks in figures 6.6 and 6.7, which promote iterating and partitioning a portion of the number line.

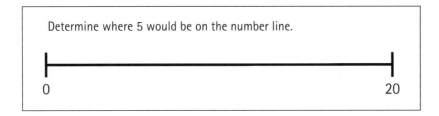

Fig. 6.6. A task involving placing 5 on the number line

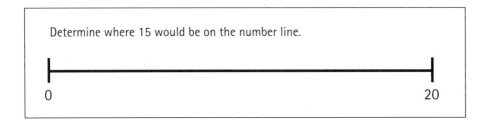

Fig. 6.7. A task involving placing 15 on the number line

Both of these number line tasks involve partitioning the number line by using a strategy of repeated halving. Further, these tasks require partitioning number lines of different lengths, an important skill that encourages students to apply multiplicative reasoning for partitioning rather than additive reasoning.

Contexts that involve measuring with different units also promote the use of multiplicative reasoning. Tasks such as those in figures 6.8 and 6.9, for example, can motivate students to discuss the relationships among units, the total quantity, and the measure or count.

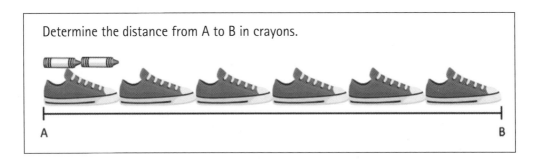

Fig. 6.8. A task involving determining the distance from A to B by using
the length of shoes and crayons

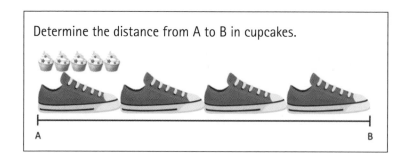

Fig. 6.9. A task involving determining the distance from A to B by using
the length of shoes and cupcakes

Determining the number of crayons needed to measure the distance from A to B
in figure 6.8 can encourage the use of multiplicative reasoning, since the length
of 2 crayons is the same as the length of each shoe. Therefore, students can reason
that the number of crayons needed to measure the distance from A to B is double
the number of shoes. The task in figure 6.9 encourages similar reasoning, since
the number of cupcakes needed to determine the length from A to B is 5 times the
number of shoes. Such multiplicative reasoning can lay a foundation of an understanding of units and multiplicative relationships that students will encounter in
grades 3–5.

Extending Knowledge of Multiplication and Division in Grades 6–8

A look ahead to how students extend ideas that they develop in grades 3–5 when
they move on to grades 6–8 reveals that students' multiplicative reasoning and understanding of multiplication serve as the foundation for many mathematical topics
that they encounter in these later grades. For example, students must apply and
extend their reasoning with fractions, ratios, proportions, measurement, probability, statistics, and algebra. Students who fail to reason multiplicatively are likely to
have a superficial understanding of many of these topics.

Many ideas related to fractions are introduced in grades 3–5 and extended in
grades 6–8, as a look at the use of fractions, ratios, and proportions in the later
grades demonstrates. As Chval, Lannin, and Jones (2013) demonstrate in *Putting
Essential Understanding of Fractions into Practice in Grades 3–5*, to reason about
fractions, students need to apply ideas related to iterating and partitioning, while
recognizing the role of the unit. Examine the problem shown in figure 6.10. Consider how you could draw a diagram to represent this situation and solve the problem.

Mary ran 12 miles this morning. The distance that Mary ran is $^3/_2$ the distance that Sarah ran. How far did Sarah run?

Fig. 6.10. Running problem

Solving the Running problem requires viewing the 12 miles that Mary ran as $^3/_2$ of the "whole," or the multiplicative unit representing the distance that Sarah ran. Determining the distance that Sarah ran involves recognizing that $^3/_2$ means 3 parts (or iterations), each of which is $^1/_2$ of that whole, or the distance that Sarah ran. Therefore, we would partition the 12 miles into 3 equal-sized parts (each 4 miles in length). Each of these 3 parts would represent $^1/_2$ of the distance that Sarah ran. We could then iterate this distance (4 miles) twice to determine the distance of Sarah's run (8 miles). The diagram in figure 6.11 represents the situation.

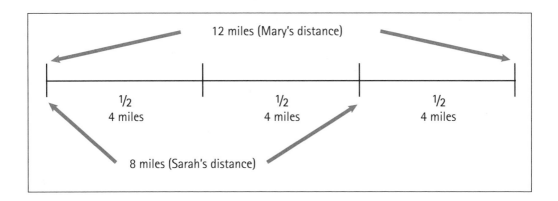

Fig. 6.11. A diagram representing the distance that Mary and Sarah ran

As the diagram and the associated thinking indicate, students must attend to the unit and use the meaning of fractions to iterate and partition the length to determine how far Sarah ran. The experiences that grades 3–5 students have with identifying the multiplicative unit as well as iterating and partitioning as they work with models of multiplication and division provide the foundational understanding that are essential for their development of a deep understanding of the use of fractions in later grades. Note, too, that this situation could be represented symbolically by $^3/_2 \times$? = 12 or 12 ÷ $^3/_2$ = ?. This context involving a partitive division with fractions connects with the partitive models for division that students encounter with whole numbers in grades 3–5.

As noted by Lamon (1999), approximately half the adult population demonstrates difficulty in reasoning proportionally, underscoring the challenges that teachers face in developing in students a deep understanding of ratio and the use of multiplicative reasoning. Students in grades 3–5 initially develop an understanding of multiplication with varying types of units. As they work with fractions, they identify a unit and carefully attend to that unit as they solve problems by applying additive or multiplicative reasoning. As students work with ratio and proportion, they compare situations with different wholes or units. For example, given the data in the table in figure 6.12, they might be asked to determine which company produces the most reliable light bulbs. The work with multiplication and division that students do in grades 3–5 lays a foundation that enables them to engage in reasoning that involves different units in grades 6–8.

	Number of defective light bulbs	Total number of light bulbs tested
Company A	10	120
Company B	22	500
Company C	7	50

Fig. 6.12. A table showing the number of defective light bulbs for three companies

An additive comparison would result in incorrect conclusions in determining the company with the most reliable light bulbs. Students who reasoned additively, and not multiplicatively, would choose company C, despite the fact that this company has the highest percentage of defective light bulbs. Students who reasoned multiplicatively would compare the ratio of defective bulbs with the total number of light bulbs tested and recognize that company B has the lowest percentage of defective bulbs.

Multiplying and dividing fractions

The understanding of multiplication that students develop in grades 3–5 is essential to the development of their understanding of multiplying and dividing fractions in grades 6–8. To examine students' development of understanding of multiplication with fractions, consider the contextual situation for $3/4 \times 2/3$ shown in figure 6.13.

Jennifer is making pancakes. Her recipe calls for $^2/_3$ of a cup of oil for each batch. She would like to make $^3/_4$ of a batch of pancakes. How much oil will she need to use?

Fig. 6.13. A contextual situation for multiplying fractions

In this contextual situation, the multiplicative unit is $^2/_3$ of a cup of oil, as represented in figure 6.14. The large rectangle represents 1 cup of oil, and the multiplicative unit of $^2/_3$ of a cup has been shaded.

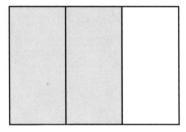

Fig. 6.14. The multiplicative unit of 2/3 of a cup of oil

Students need to interpret the $^3/_4$ in this situation as the multiplier, scalar, or operator (Barnett-Clarke et al. 2010). Under each view, determining how much oil Jennifer will need for her pancakes involves applying $^3/_4$ to the multiplicative unit, $^2/_3$. Students can do this by first partitioning the multiplicative unit into 4 equal parts. This process is illustrated in the diagram in figure 6.15, where each of the four equal parts is represented by two small shaded rectangles. The dashed lines in the diagram make it easy to identify each small rectangle as $^1/_{12}$ of a cup of oil.

Fig. 6.15. The multiplicative unit of $^2/_3$ of a cup of oil, partitioned into 4 equal parts

A pair of small shaded rectangles represents $^1/_4$ of the multiplicative unit. Because in the pancake situation Jennifer is interested in $^3/_4$ of the multiplicative unit, students need to consider 3 pairs, or 6, of the small shaded rectangles, as identified by stars in figure 6.16. Each of the 6 small shaded rectangles identified by stars represents $^1/_{12}$ of a cup of oil. Therefore, $^3/_4$ of $^2/_3$ of a cup of oil is $^6/_{12}$, or $^1/_2$, of a cup of oil. Symbolically, $^3/_4 \times {}^2/_3 = {}^6/_{12} = {}^1/_2$.

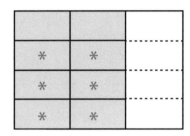

Fig. 6.16. $^3/_4$ of the multiplicative unit of $^2/_3$ of a cup of oil, identified by stars

Use of the array model can help give meaning to the common algorithm for multiplying fractions. Notice that the denominator of the product (12) is the product of the denominators of the multiplier (4) and the multiplicand (3). Students can make sense of this result when they examine the array model, constructed by partitioning a large rectangle first into 3 equal columns and then into four equal rows, giving a total of 12 identical small rectangles. Similarly, the numerator in the product (6) is the product of the numerators of the multiplier (3) and the multiplicand (2). Again, students can make sense of this result by examining the array model. The 6 was obtained in the model by iterating and shading 2 thirds, and then marking 3 pairs of small rectangles, for a total of 6 small rectangles to consider in the context of Jennifer's pancakes.

Thus, a model can illustrate multiplication of fractions. By understanding the model, students will be able to develop an understanding of the written algorithm for multiplying fractions—understanding that is essential to their mathematical development in grades 6–8. As Otto and colleagues (2011, p. 61) show, this algorithm may also be justified through the use of the commutative and associative properties of multiplication.

Finally, consider students' understanding of $^2/_3 \times {}^3/_4$ in relation to $^3/_4 \times {}^2/_3$. The commutative property guarantees that these products are the same. If students compare diagrams for $^3/_4 \times {}^2/_3$ and $^2/_3 \times {}^3/_4$ such as those in figure 6.16 and figure 6.17, respectively, however, they will see that the order of the factors may permit determining one product with greater ease than another. In this particular example, it is easier to "see" $2/3$ of $3/4$ than $3/4$ of $2/3$ because $3/4$ is immediately partitioned into three equal parts in the number line model representing $2/3 \times 3/4$.

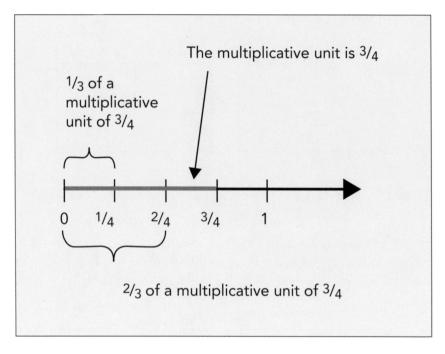

Fig 6.17. Illustrating $2/3 \times 3/4$ on a number line

In the same way that the understanding of multiplication that students develop in grades 3–5 underpins the understanding of multiplication of fractions that they build in grades 6–8, the understanding of division that students acquire in grades 3–5 supplies the foundation for their later development of understanding of division with fractions. Consider the contextual situation for $1\frac{1}{2} \div 1/3$, shown in figure 6.18.

Elliot is making chocolate cupcakes. Each batch calls for $1/3$ of a cup of cocoa. He has $1\frac{1}{2}$ cups of cocoa. How many batches (full and partial) of chocolate cupcakes can he make?

Fig. 6.18. A contextual situation for division of fractions

This is a measurement division situation, in which $1/3$ of a cup of cocoa represents the multiplicative unit, and the task is to determine the number of multiplicative units contained in $1\frac{1}{2}$ cups of cocoa. To represent this situation, students might begin with the dividend, which is $1\frac{1}{2}$ cups of cocoa. If they used one rectangle to represent 1 cup of cocoa, they might draw two rectangles and shade 1 entire rectangle and $\frac{1}{2}$ of the second rectangle, as shown in figure 6.19, to represent the entire amount of cocoa that Elliot has.

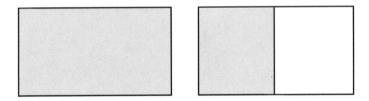

Fig. 6.19. The dividend, $1\frac{1}{2}$ cups of cocoa, represented in an area model

They need to find the number of $\frac{1}{3}$ cups of cocoa in $1\frac{1}{2}$ cups. They could use horizontal lines to partition these rectangles into thirds. From the whole cup, they could get a total of 3 of these multiplicative units, as shown in figure 6.20.

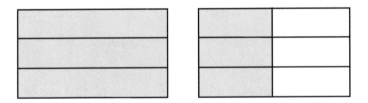

Fig. 6.20. Partitioning $1\frac{1}{2}$ cups of cocoa into groups of $\frac{1}{3}$ cup

They could rearrange the shaded portion of the rectangle on the right, as in figure 6.21, to see that Elliot has another $\frac{1}{3}$ of a cup of cocoa (the multiplicative unit) in the $\frac{1}{2}$ cup of cocoa. The figure shows a total of 4 of these $\frac{1}{3}$ cups of cocoa, along with a remainder. Using this model, students could easily see that the remainder is $\frac{1}{6}$ of a cup of cocoa, and that this is also equal to $\frac{1}{2}$ of the multiplicative unit of $\frac{1}{3}$ of a cup of cocoa. Thus, they would be able to say that $4\frac{1}{2}$ multiplicative units are contained in $1\frac{1}{2}$ cups of cocoa, so $1\frac{1}{2} \div \frac{1}{3} = 4\frac{1}{2}$. In the context of the problem, they would be able to understand that Elliot has enough cocoa to make $4\frac{1}{2}$ batches of chocolate cupcakes.

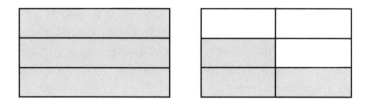

Fig. 6.21. Identifying the number of $1/3$ cups of cocoa in $1^{1}/_{2}$ cups

Writing and manipulating algebraic expressions

An important connection exists between the understanding of multiplication and division that students develop in grades 3–5 and the algebraic reasoning that they must do in grades 6–8. Many of the multiplicative ideas that students use in grades 3–5 allow them to write, make sense of, and manipulate algebraic expressions of increasing sophistication. For example, students might be asked to write an expressions for the cost of n pencils, given that each pencil costs 25¢. As students write the expression $25n$, they identify the multiplicative unit as 25 and the number of multiplicative units as n. The expression $25n$ represents the product, the multiplicative unit, and the number of multiplicative units simultaneously. Furthermore, students apply the commutative property to their use of this notation. They should recognize that $n \times 25 = 25 \times n = 25n$, gradually learning that the standard form for writing multiplication with a variable is to write the constant first, and that they can write $25n$ in place of $n25$ because the commutative property of multiplication applies to all real numbers and n and 25 are both real numbers.

In studying algebra, students use and extend the properties discussed in Chapter 5 to the real numbers and to varying unknown quantities. For example, students apply the distributive property when, by combining like terms, they write $5m + 4m = 9m$. The often unwritten support for this statement is that $5m + 4m = (5 + 4)m = 9m$. The associative property is used to rewrite the expression $7(3y)$ as $21y$. Students also apply multiplication and division by 0 and 1 in many instances—for example, solving equations such as $x(x + 3) = 0$ by noting that either $x = 0$ or $x + 3 = 0$.

Another connection from students' work with multiplication and division in grades 3–5 to their later work in algebra involves the use of the partial products algorithm that was introduced in Chapter 5. In grades 6–8, as students consider multiplying various expressions, they are introduced to what appears to be a new and problematic situation—multiplying two binomials, such as $(n + 3)(n + 4)$. However, students should have had many opportunities to represent and model these types of situations when they multiplied 13 and 14 during their introduction to multi-digit multiplication. As the model for $(n + 4)$ $(n + 3)$ in figure 6.22 illustrates, algebra students can create models that identify each of the partial products, similar to the partial products calculated in multiplication of whole numbers. The partial products algorithm allows students to connect multiplication with whole numbers readily with the multiplication of binomials.

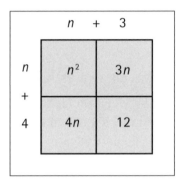

Fig. 6.22. A model of $(n + 4)$ $(n + 3)$

Conclusion

Just a few examples from the content in grades 6–8 show that the foundational understanding of multiplication and division that students develop in grades 3–5 is essential for the learning that follows. For instance, the work with multiplication and division in grades 3–5 provides an entry point for reasoning about various mathematical topics such as ratio, rate, proportion, and percentage. Moreover, this work influences student reasoning in mathematical strands, including geometry, probability, statistics, and algebra. Considerable time and effort are needed to develop an understanding in grades 3–5 that will be sufficiently deep to ensure that students will be successful in grades 6–8. As a teacher, you play a critical role in developing such understanding through your careful selection of instructional strategies and assessments, use of your students' current and prior understanding, and knowledge of curriculum.

Appendix 1
The Big Ideas and Essential Understandings for Multiplication and Division

This book focuses on essential understandings that are identified and discussed in *Developing Essential Understanding of Multiplication and Division for Teaching Mathematics in Grades 3–5* (Otto et al. 2011). For the reader's convenience, the full list of the big ideas and essential understandings in that book is reproduced below. The big idea and essential understandings that are the special focus of this book are highlighted in orange.

Big Idea 1. Multiplication is one of two fundamental operations, along with addition, which can be defined so that it is an appropriate choice for representing and solving problems in many different situations.

Essential Understanding 1*a*. In the multiplicative expression $A \times B$, A can be defined as a *scaling factor*.

Essential Understanding 1*b*. Each multiplicative expression developed in the context of a problem situation has an accompanying explanation, and different representations and ways of reasoning about a situation can lead to different expressions and equations.

Essential Understanding 1*c*. A situation that can be represented by multiplication has an element that represents the scalar and an element that represents the quantity to which the scalar applies.

Essential Understanding 1*d*. A scalar definition of multiplication is useful in representing and solving problems beyond whole number multiplication and division.

Essential Understanding 1*e*. Division is defined by its inverse relationship with multiplication.

Essential Understanding 1*f*. Using proper terminology and understanding the division algorithm provide the basis for understanding how numbers such as the *quotient* and the *remainder* are used in a division situation.

Big Idea 2. The properties of multiplication and addition provide the mathematical foundation for understanding computational procedures for multiplication and division, including mental computation and estimation strategies, invented algorithms, and standard algorithms.

Essential Understanding 2*a*. The commutative and associative properties of multiplication and the distributive property of multiplication over addition ensure flexibility in computations with whole numbers and provide justifications for sequences of computations with them.

Essential Understanding 2*b*. The right distributive property of division over addition allows computing flexibly and justifying computations with whole numbers, but there is no left distributive property of division over addition and no commutative or associative property of division of whole numbers.

Essential Understanding 2*c*. *Order of operations* is a set of conventions that eliminates ambiguity in, and multiple values for, numerical expressions involving multiple operations.

Essential Understanding 2*d*. Properties of operations on whole numbers justify written and mental computational algorithms, standard and invented.

Appendix 2
Resources for Teachers

The following list highlights a few of the many books, articles, videos, and websites that are helpful resources for teaching multiplication and division in grades 3–5. Abstracts from the publishers provide brief descriptions of the resources.

Books

Battista, Michael T. *Cognition-Based Assessment & Teaching of Multiplication and Division: Building on Students' Reasoning.* Portsmouth, N.H.: Heinemann, 2012.

> Using a research-based framework that describes the development of students' thinking and learning in terms of levels of sophistication, a "cognitive terrain" that includes ascents and plateaus, Battista shows how teachers can build on their students' reasoning.

Carpenter, Thomas, Elizabeth Fennema, Megan Loef Franke, Linda Levi, and Susan B. Empson. *Children's Mathematics/Cognitively Guided Instruction.* Portsmouth, N.H.: Heinemann, 1999.

> By the time children begin school, most have already developed a sophisticated, informal understanding of basic mathematical concepts and problem-solving strategies. Too often, however, the mathematics that classroom instruction imposes on them fails to connect with this informal knowledge. The authors developed this book to help teachers understand children's intuitive mathematical thinking and use that knowledge to help them learn mathematics with understanding. Based on more than twenty years of research, the book portrays the development of children's understanding of basic number concepts. The authors offer a detailed explanation and numerous examples of the problem-solving and computational processes that virtually all children use as their numerical thinking develops. They also describe how classrooms can be organized to foster that development. Two accompanying CDs include videos that provide an inside look at students and teachers in real classrooms implementing the teaching and learning strategies described in the text.

Fosnot, Catherine Twomey. *Exploring Parks and Playgrounds: Multiplication and Division of Fractions.* Portsmouth, N.H.: Heinemann, 2008.

This title is one of five units in *Contexts for Learning Mathematics: Investigating Fractions, Decimals, and Percents* (4–6). The focus of this unit is the development of students' understanding of multiplication with rational numbers. The unit uses the context of parks and playgrounds to introduce the double number line and the array models as helpful tools. Initially, students explore the situation of two cousins running on a 26-mile course that winds through a park. In this first investigation, the distributive property and the use of landmark fractions (such as $1/2$) provide a focus.

————. *Groceries, Stamps, and Measuring Strips: Early Multiplication*. Portsmouth, N.H.: Heinemann, 2008.

This title is one of five units in *Contexts for Learning Mathematics: Investigating Multiplication and Division* (3–5). The focus of this unit is the introduction and early development of multiplication. By making use of realistic contexts, the unit invites students to find ways to mathematize their lived worlds with grouping structures. The unit uses many contexts: inside the grocery store; postage stamps; city buildings, windows, and buses; tiled patios; a baker's trays; and sticker pages.

————. *Muffles' Truffles: Multiplication and Division with the Array*. Portsmouth, N.H.: Heinemann, 2008.

This title is one of five units in *Contexts for Learning Mathematics: Investigating Multiplication and Division* (3–5). The focus of this unit is the development of the open array as a model for multiplication and division. The unit uses a series of investigations based on the context of a candy shop called Muffles' Truffles. The questions posed in the first investigation (how many boxes of ten can be made with a given quantity of truffles; how many leftovers will there be from a given quantity, and how can they be combined to make assortment boxes; and what is the cost of a given quantity of truffles if they cost $1 each?) give students an opportunity to explore place value, the multiplicative structure of our base-ten system, and measurement division. In the second and third investigations, students build two-dimensional blueprints of one-layer boxes and use these arrays to explore some of the big ideas in multiplication (the distributive, associative, and commutative properties). In the fourth and final investigation, students work with open arrays in the context of labeling and pricing wrapped boxes of truffles.

————. *The Big Dinner: Multiplication with the Ratio Table*. Portsmouth, N.H.: Heinemann, 2008.

The preparation of a turkey dinner provides a context for introducing early multiplication strategies and supports automatizing the facts, using the ratio table, and developing the distributive property with large numbers. Strings of problems guide learners toward computational fluency with whole-number multiplication and build automaticity with multiplication facts by focusing on relationships.

————. *The Box Factory: Extending Multiplication with the Array*. Portsmouth, N.H.: Heinemann, 2008.

This title is one of five units in *Contexts for Learning Mathematics: Investigating Multiplication and Division* (3–5). The focus of this unit is the deepening and extending students' understanding of multiplication, specifically the associative and commutative properties and their use in computation, systematic factoring, and the extension of students' understanding of two-dimensional rectangular arrays to three-dimensional arrays within rectangular prisms. The unit includes a series of investigations set in the context of a cardboard box factory. Initially students design a variety of boxes (rectangular prisms) that hold 24 items arranged in rows, columns, and layers. The questions posed in the first investigation (how many box arrangements are possible, and how do we know for certain that we have found all the possibilities?) give students opportunities to explore the associative and commutative properties, factor pairs, doubling and halving strategies, and systematic ways of organizing their work to determine all possible cases. In the second and third investigations, students analyze the amount and cost of the cardboard needed for their boxes, deepening their understanding of the associative property, examining congruency versus equivalency, and exploring the relationship of the surface area to the shape of the box. Subsequent investigations involve using two different cubic boxes as units of measurement, and determining the volume of a shipping box that measures 4 feet by 6 feet by 4 feet. By the end of the unit, formulas for surface area and volume of rectangular prisms are the focus.

————. *The Teachers' Lounge: Place Value and Division.* Portsmouth, N.H.: Heinemann, 2008.

This title is one of five units in *Contexts for Learning Mathematics: Investigating Multiplication and Division* (3–5). The focus of this unit is division. It begins with the story of a teacher noticing a service person fill two different vending machines in the teachers' lounge with beverages. This story sets the stage for a series of investigations designed to support students' development of a repertoire of strategies for multiplication and division, including the use of—

- the ten-times strategy;
- partial products and partial quotients;
- the associative property;
- the distributive property of multiplication over addition.

Fosnot, Catherine Twomey, and Maarten Dolk. *Young Mathematicians at Work: Constructing Multiplication and Division.* Portsmouth, N.H.: Heinemann, 2002.

Fosnot and Dolk focus on how to develop an understanding of multiplication and division in grades 3-5. Their book accomplishes the following:

- Provides strategies to help teachers turn their classrooms into math workshops that encourage and reflect mathematizing
- Examines several ways to engage and support children as they construct important strategies and big ideas related to multiplication
- Takes a close look at the strategies and big ideas related to division
- Defines modeling and provides examples of how learners construct models—with a discussion of the importance of context
- Discusses what it means to calculate by using number sense and whether algorithms should still be the goal of computation instruction
- Describes how to strengthen performance and portfolio assessment

Fosnot, Catherine Twomey, and William Uttenbogaard. *Minilessons for Early Multiplication and Division: A Yearlong Resource.* Portsmouth, N.H.: Heinemann, 2008.

This guide offers 75 mini-lessons to choose from throughout the year. In contrast with investigations, which are the heart of the math workshop, the mini-lesson is more guided and more explicit, designed for use at the start of the workshop and lasting for ten to fifteen minutes. Mini-lessons in this resource provide students with experiences to develop efficient computation and can also be used with small groups of students to differentiate instruction.

————. *Minilessons for Extending Multiplication and Division: A Yearlong Resource.* Portsmouth, N.H.: Heinemann, 2008.

This guide contains 77 mini-lessons structured as strings of related computation problems. They are designed to generate discussion of certain strategies or big ideas that are markers on the landscape of learning for multiplication and division, particularly using numbers with two and three digits. Although the emphasis is on the development of mental arithmetic strategies, learners do not have to solve the problems in their heads. Students should be encouraged to examine the numbers in each problem and think about clever, efficient ways to solve it.

Harel, Guershon, and Jere Confrey. *The Development of Multiplicative Reasoning in the Learning of Mathematics.* Albany: State University of New York Press, 1994.

This book includes research on the development of multiplicative concepts.

Otto, Albert Dean, Janet H. Caldwell, Cheryl Ann Lubinski, and Sarah Wallus Hancock. *Developing Essential Understanding of Multiplication and Division for Teaching Mathematics in Grades 3–5.* Essential Understanding Series. Reston, Va.: National Council of Teachers of Mathematics, 2011.

"Unpacking" the ideas related to multiplication and division is a critical step in developing a deeper understanding. To those without specialized training, many of these ideas might appear to be easy to teach. But those who teach in grades 3–5 are aware of their subtleties and complexities. This book identifies and examines two big ideas and related essential understandings for teaching multiplication and division in grades 3–5. The authors examine the ways in which counting, adding, and subtracting lead to multiplication and division, as well as the role that these operations play in algebraic expressions and other advanced topics. The book examines challenges in teaching, learning, and assessment and is interspersed with questions for teachers' reflection.

Van de Walle, John, Karen S. Karp, and Jennifer M. Bay-Williams. *Elementary and Middle School Mathematics: Teaching Developmentally.* 7th ed. Allyn & Bacon, 2010.

The authors wrote this book to help teachers understand mathematics and become confident in their ability to teach the subject to children in kindergarten through eighth grade. The chapters related to teaching and learning of multiplication and division provide ideas and insights that can support teachers as they design and implement their lessons.

Articles

Azim, Diane. "Understanding Multiplication as One Operation." *Mathematics Teaching in the Middle School* 7 (April 2002): 466–71.

> The author introduces two teaching methods and two student-created methods for conceptualizing multiplication with rational numbers.

Drake, Jill Mizell, and Angela T. Barlow. "Assessing Students' Levels of Understanding Multiplication through Problem Writing." *Teaching Children Mathematics* 14 (December 2007/January 2008): 272–77.

> The authors discuss the potential of problem writing as a technique for assessing the depth of students' mathematical understandings. Discussions include samples of student-generated word problems and make connections between problem writing and the NCTM Process Standards.

Fuson, Karen C. "Toward Computational Fluency in Multidigit Multiplication and Division." *Teaching Children Mathematics* 9 (February 2003): 300–305.

> The author introduces an alternative to traditional instruction in multiplication and division to develop students' computational fluency.

Gregg, Jeff, and Diana Underwood Gregg. "Interpreting the Standard Division Algorithm in a 'Candy Factory' Context." *Teaching Children Mathematics* 14 (August 2007): 25–31.

> The authors discuss difficulties that preservice teachers experience when they try to make sense of the standard division algorithm. They also describe a realistic context that can be productive in helping students think about why the algorithm works and the role that place value has in it. Real-world connections and student work bring meaning to the problems in the elementary classroom.

Hedges, Melissa, DeAnn Huinker, and Meghan Steinmeyer. "Unpacking Division to Build Teachers' Mathematical Knowledge." *Teaching Children Mathematics* 11 (May 2005): 478–83.

> The authors examine preservice teachers' understanding of division.

Lamberg, Teruni, and Lynda R. Wiest. "Conceptualizing Division with Remainders." *Teaching Children Mathematics* 18 (March 2012): 426–33.

> The authors share third graders' struggles to solve contextualized, student-generated division problems with remainders.

Lin, Cheng-Yao. "Teaching Multiplication Algorithms from Other Cultures." *Teaching Mathematics in the Middle School* 13 (December 2007): 298–304.

> The authors share multiplication algorithms from different world cultures, including Hindu, Egyptian, Russian, Japanese, and Chinese. Students can learn these algorithms and gain a better understanding of the operation and properties of multiplication. Tabular and pictorial demonstrations of the algorithms are included.

Martin, John F., Jr. "The Goal of Long Division." *Teaching Children Mathematics* 15 (April 2009): 482–87.

> The author describes thinking that students need to develop to facilitate their understanding of answers resulting from division. Keys to this understanding include the role of place value in the quotient and the power of multiples of ten in determining the quotient.

Matteson, Shirley M. "A Different Perspective on the Multiplication Chart." *Mathematics Teaching in the Middle School* 16 (May 2011): 562–66.

> The author introduces the use of a "multiples mat" to describe patterns, solve addition and subtraction problems, and explore fractions, ratios, proportions, and sample functions.

Nugent, Patricia M. "Lattice Multiplication in a Preservice Classroom." *Mathematics Teaching in the Middle School* 13 (September 2007): 110–13.

> The author discusses the development of preservice students' conceptual understanding of the lattice multiplication algorithm. The conceptualization of the algorithm is extended from the multiplication of whole numbers to the multiplication of polynomials.

Roberts, Sally K. "Snack Math: Young Children Explore Division." *Teaching Children Mathematics* 9 (January 2003): 258–61.

> The author introduces an exploration of young children's intuitive strategies for partitioning sets of objects.

Turner, Erin E., Debra L. Junk, and Susan B. Empson. "The Power of Paper-Folding Tasks: Supporting Multiplicative Thinking and Rich Mathematical Discussion." *Teaching Children Mathematics* 13 (February 2007): 322–29.

> The authors focus on paper folding and describe its potential to support the development of multiplicative thinking and facilitate discussion of mathematical problems for all students.

Wickett, Maryann S. "Discussion as a Vehicle for Demonstrating Computational Fluency in Multiplication." *Teaching Children Mathematics* 9 (February 2003): 318–21.

> To help children make sense of multiplication, the author introduces an activity called "Silent Multiplication," which encourages students to use what they know about easier multiplication problems to solve increasingly difficult, related problems mentally.

Videos

Integrating Mathematics and Pedagogy (IMAP)
http://www.sci.sdsu.edu/CRMSE/IMAP/video.html

> *IMAP: Select Videos of Children's Reasoning* is a CD containing twenty-five video clips of elementary school children engaged in mathematical thinking. The CD runs on PC and Mac platforms and comes with an interface that includes the

transcript (full or synchronized) and background information for each clip. Also included on the CD is a video guide containing questions to consider before and after viewing each video clip, interviews that teachers or prospective teachers can use when working with children, and other resources.

See also Carpenter and colleagues (1999), under "Books."

> *Children's Mathematics/Cognitively Guided Instruction* has two accompanying CDs that provide an inside look at students and teachers in real classrooms implementing the teaching and learning strategies described in the text.

Online Resources

NCTM Illuminations Lessons
http://illuminations.nctm.org/Lessons.aspx

> A project of the National Council of Teachers of Mathematics, Illuminations is part of the Verizon Thinkfinity program. The Illuminations website offers hundreds of standards-based lessons. Select the types of lessons that you want, as well as the appropriate grade band, and click "Search."

Illustrative Mathematics Project
http://www.illustrativemathematics.org

> Illustrative Mathematics provides guidance to states, assessment consortia, testing companies, and curriculum developers by illustrating the range and types of mathematical work that support implementation of the Common Core State Standards. One tool on this website is a growing collection of mathematical tasks that are organized by standard for each grade level and illustrate important features of the indicated standard or standards. The tasks on the website are not meant to be considered in isolation. Taken together in sets, these tasks are intended to illustrate a particular standard. Eventually, the site will showcase sets of tasks for each standard that—
>
> - illuminate the central meaning of the standard and also show connections with other standards;
> - clarify what is familiar about the standard and what is new with the advent of the Common Core State Standards;
> - include both teaching and assessment tasks; and
> - reflect the full range of difficulty that the standard expects students to master.

Appendix 3
Tasks

This book examines rich tasks that have been used in the classroom to bring to the surface students' understandings and misunderstandings about multiplication and division. These tasks are reproduced here, in the order in which they appear in the book, for the reader's personal or classroom use.

Task 1

1. Draw a diagram of 7 groups of 3 circles.

2. Draw a diagram to represent 7 × 3 inside the box.

3. What is 7 × 3? _____

Task 2

Morris and Leslie both made pictures of the sky at night.
Morris used 60 star stickers for his picture.
Leslie used 20 star stickers for her picture.

Anna said, "Morris used 3 times as many star stickers as Leslie."
Is Anna correct?   Circle your answer: Yes   No

Explain your thinking.

Ben said, "Leslie used 3 times as many star stickers as Morris."
Is Ben correct?   Circle your answer: Yes   No

Explain your thinking.

Cade said, "Leslie used 40 more stickers than Morris."
Is Cade correct?   Circle your answer: Yes   No

Explain your thinking

Task 3

Marlene made 6 batches of muffins. There were 24 muffins in each batch. Which of the following number sentences could be used to find the number of muffins that she made?

Circle the correct number sentence below.

6 × _____ = 24

6 + 24 = _____

6 + _____ = 24

6 × 24 = _____

Explain your thinking.

Task 4

Write a word problem that you could solve by computing 7 × 9.

Task 5

Maria has 5 bags of oranges. Each bag has 7 oranges.
How many oranges does she have?

Show how you thought about the problem:

Number of oranges:

Task 6

The following picture shows a floor in a room that is covered with square tiles. A rug is placed over some of the tiles. How many tiles are on the floor in this room?

Number of tiles on the floor _____

How did you determine the number of tiles?

Task 7

20 ÷ 4 = ?

Write a word problem that you would solve by computing 20 ÷ 4.

Make a diagram to show 20 ÷ 4 in your problem.

Task 8

Show your work on the following tasks:

Balloons

A mother had 20 balloons. She wanted to give them to her 3 children so that each child would have the same number of balloons. How many balloons did each child get?

Birds

A pet store owner has 14 birds and some cages. She will put 3 birds in each cage. How many cages will she need?

Cookies

A father has 17 cookies. He wants to give them to his 3 children so that each child has the same number of cookies. How many cookies will each child get?

Task 9

Kara says that when you multiply two numbers, the answer is always bigger.

Do you agree with Kara?   Circle one: Yes   No

Explain your thinking.

Task 10

Situation A

Seth has 37 boxes of marbles with 28 marbles in each box. Marcus has 28 boxes of marbles with 37 marbles in each box. Who has more marbles?

Circle one: Seth has more.   Marcus has more.   They have the same
                                                                                number.

Explain your thinking.

Situation B

Which is more? Circle one: 37 × 28   or   28 × 37

Explain your thinking.

Task 11

50 × 40 is 2,000. What is an easy way to find 49 × 40?

Explain your easy way.

Task 12

Determine the missing numbers in the following tasks. Explain your thinking for each part.

A. 18 × 6 is the same as _____ × 12

B. 24 × 12 is the same as 6 × _____

C. 200 ÷ 6 is the same as 100 ÷ _____

D. 200 ÷ 6 is the same as _____ ÷ 12

Solve the following "repackaging" problems. Explain your thinking in each case.

1. Rosa has 18 bags with 6 marbles in each bag. She wants to repackage the marbles with 12 marbles in each bag. How many bags will Rosa need?

2. Lucy has 24 bags with 12 marbles in each bag. She wants to repackage the marbles so that she uses 6 bags and she has the same number of marbles in each bag. How many marbles will Lucy have in each bag?

Task 13

Mike said, "721 ÷ 7 is the same as 700 ÷ 7 + 21 ÷ 7."

Do you agree with Mike?   Circle one: Yes   No

Why or why not?

Task 14

(a) What is the answer to 5 + 5 + 5 × 0?

(b) Patrick says that 6 + 5 × 0 is equal to 0 because 11 × 0 = 0.

Courtney says that 6 + 5 × 0 is equal to 6. She said, "I took 5 × 0 first, which is 0. Then I took 6 + 0, so the answer is 6."

Who is correct?   Circle one: Patrick   Courtney

Task 15

Marsha is in the middle of multiplying 37 × 54. So far she has written this:

$$
\begin{array}{r}
{}^{2} \\
37 \\
\times\ \underline{54} \\
148 \\
0
\end{array}
$$

(a) What does Marsha's little "2" above the 3 in 37 mean?

(b) Why has Marsha put a "0" below 148?

References

Ameis, Jerry A. "The Truth about PEDMAS." *Mathematics Teaching in the Middle School* 16 (March 2011): 414–20.

Ball, Deborah Loewenberg. "Magical Hopes: Manipulatives and the Reform of Math Education." *American Educator: The Professional Journal of the American Federation of Teachers* 16 (Summer 1992): 14–18.

Barnett-Clarke, Carne, William Fisher, Rick Marks, and Sharon Ross. *Developing Essential Understanding of Rational Numbers for Teaching Mathematics in Grades 3–5.* Essential Understanding Series. Reston, Va.: National Council of Teachers of Mathematics, 2010.

Bass, Hyman. "Computational Fluency, Algorithms, and Mathematical Proficiency: One Mathematician's Perspective." *Teaching Children Mathematics* 9 (February 2003): 322–27.

Battista, Michael T. *Cognition-Based Assessment & Teaching of Fractions: Building on Students' Reasoning.* Portsmouth, N.H.: Heinemann, 2012.

Buswell, Guy T. *Diagnostic Studies in Arithmetic.* Chicago: University of Chicago, 1926.

Carpenter, Thomas P., Elizabeth Fennema, and Megan L. Franke. "Cognitively Guided Instruction: A Knowledge Base for Reform in Primary Mathematics Instruction." *Elementary School Journal* 97 (September 1996): 3–20.

Chval, Kathryn B, and Óscar Chávez. "Designing Math Lessons for English Language Learners." *Mathematics Teaching in the Middle School* (December 2011/January 2012): 261–65.

Chval, Kathryn B., Óscar Chávez, Sarah Pomerenke, and Kari Reams. "Enhancing Mathematics Lessons to Support All Students." In *Mathematics for Every Student: Responding to Diversity, Grades Pre-K–5,* edited by Dorothy Y. White and Julie Sliva Spitzer, pp. 43–52. Reston, Va.: National Council of Teachers of Mathematics, 2009.

Chval, Kathryn B., and Jane Davis. "The Gifted Student." *Mathematics Teaching in the Middle School* 14 (December 2008/January 2009): 267–74.

Chval, Kathryn B., John Lannin, and Dusty Jones. *Putting Essential Understanding of Fractions into Practice in Grades 3–5.* Putting Essential Understanding into Practice Series. Reston, Va.: National Council of Teachers of Mathematics, 2013.

Clarke, Barbara, and Doug Clarke. "Mathematics Teaching in Grades K–2: Painting a Picture of Challenging, Supportive, and Effective Classrooms." In *Perspectives on the Teaching of Mathematics,* Sixty-sixth Yearbook of the National Council of Teachers of Mathematics (NCTM), edited by Rheta N. Rubenstein, pp. 67–81. Reston, Va.: NCTM, 2004.

Confrey, Jere. "Splitting, Similarity, and Rate of Change: A New Approach to Multiplication and Exponential Functions." In *The Development of Multiplicative Reasoning in the Learning of Mathematics*, edited by Guershon Harel and Jere Confrey, pp. 291–330. Albany: State University of New York Press, 1994.

Confrey, Jere, and Guershon Harel. "Introduction." In *The Development of Multiplicative Reasoning in the Learning of Mathematics*, edited by Guershon Harel and Jere Confrey, pp. vii–xxviii. Albany: State University of New York Press, 1994.

Dougherty, Barbara J. "Access to Algebra: A Process Approach." In *The Future of the Teaching and Learning of Algebra*, edited by Helen Chick, Kaye Stacey, Jill Vincent, and John Vincent, pp. 207–13. Victoria, Australia: University of Melbourne, 2001.

Fosnot, Catherine Twomey, and Maarten Dolk. *Young Mathematicians at Work: Constructing Fractions, Decimals, and Percents.* Heinemann, 2002.

Fuson, Karen C. "Developing Mathematical Power in Whole Number Operations." In *A Research Companion to "Principles and Standards for School Mathematics,"* edited by Jeremy Kilpatrick, W. Gary Martin, and Deborah Schifter, pp. 68–94. Reston, Va.: National Council of Teachers of Mathematics, 2003a.

————. "Toward Computational Fluency in Multidigit Multiplication and Division." *Teaching Children Mathematics* 9 (February 2003b): 300–305.

Gravemeijer, Koeno, Richard Lehrer, Bert van Oers, and Lieven Verschaffel, eds. *Symbolizing, Modeling and Tool Use in Mathematics Education.* Mathematics Education Library, vol. 30. Dordrecht, The Netherlands: Springer, 2002.

Greer, Brian. "Multiplication and Division as Models of Situations." In *Handbook of Research on Mathematics Teaching and Learning*, edited by Douglas A. Grouws, pp. 276–95. New York: Macmillan, 1992.

————. "Extending the Meaning of Multiplication and Division." In *The Development of Multiplicative Reasoning in the Learning of Mathematics*, edited by Guershon Harel and Jere Confrey, pp. 61–85. Albany: State University of New York Press, 1994.

Grossman, Pamela L. *The Making of a Teacher: Teacher Knowledge and Teacher Education.* Professional Development and Practice Series. New York: Teachers College Press, 1990.

Hill, Heather C., Brian Rowan, and Deborah Loewenberg Ball. "Effects of Teachers' Mathematical Knowledge for Teaching on Student Achievement." *American Educational Research Journal* 42 (Summer 2005): 371–406.

Jacob, Lorraine, and Sue Willis. "Recognising the Difference between Additive and Multiplicative Thinking in Young Children." In *Numeracy and Beyond: Proceedings of the Twenty-fourth Annual Conference of the Mathematics Education Research Group of Australasia Incorporated,* edited by Bob Perry, Michael Charles Mitchelmore, and Janette Bobis, vol. 2, pp. 306–13. Sydney: Mathematics Education Research Group of Australasia, 2001.

Kabiri, Mary S., and Nancy Smith. "Turning Traditional Textbook Problems into Open-Ended Problems." *Mathematics Teaching in the Middle School* 9 (November 2003): 186–92.

Kenney, Patricia Ann, and Edward A. Silver, eds. *Results from the Sixth Mathematics Assessment of the National Assessment of Educational Progress.* Reston, Va.: National Council of Teachers of Mathematics, 1997.

Kouba, Vicky L. "Children's Solution Strategies for Equivalent Set Multiplication and Division Word Problems." *Journal for Research in Mathematics Education* 20 (March 1989): 147–58.

Lamon, Susan J. *Teaching Fractions and Ratios for Understanding: Essential Content Knowledge and Instructional Strategies for Teachers.* Mahwah, N.J.: Lawrence Erlbaum, 1999.

———. "Ratio and Proportion: Cognitive Foundations in Unitizing and Norming." In *The Development of Multiplicative Reasoning in the Learning of Mathematics*, edited by Guershon Harel and Jere Confrey, pp. 89–120. Albany: State University of New York Press, 1994.

Lampert, Magdalene. "Knowing, Doing, and Teaching Multiplication." *Cognition and Instruction* 3, no. 4 (1986): 305–42.

Magnusson, Shirley, Joseph Krajcik, and Hilda Borko. "Nature, Sources, and Development of Pedagogical Content Knowledge for Science Teaching." In *Examining Pedagogical Content Knowledge*, edited by Julie Gess-Newsome and Norman G. Lederman, pp. 95–132. Dordrecht, The Netherlands: Kluwer Academic, 1999.

Marks, Genée, and Judith Mousley. "Mathematics Education and Genre: Dare We Make the Process Writing Mistake Again?" *Language and Education* 4 (1990): 117–35.

Mulligan, Joanne, and Jane Watson. "A Developmental Multimodal Model for Multiplication and Division." *Mathematics Education Research Journal* 10, no. 2 (1998): 61–86.

Mulligan, Joanne T., and Michael C. Mitchelmore. "Young Children's Intuitive Models of Multiplication and Division." *Journal for Research in Mathematics Education* 28 (May 1997): 309–30.

National Governors Association Center for Best Practices and Council of Chief State School Officers (NGA Center and CCSSO). *Common Core State Standards for Mathematics. Common Core State Standards (College- and Career-Readiness Standards and K–12 Standards in English Language Arts and Math).* Washington, D.C.: NGA Center and CCSSO, 2010. http://www.corestandards.org.

Otto, Albert Dean, Janet H. Caldwell, Cheryl Ann Lubinski, and Sarah Wallus Hancock. *Developing Essential Understanding of Multiplication and Division for Teaching Mathematics in Grades 3–5.* Essential Understanding Series. Reston, Va.: National Council of Teachers of Mathematics, 2011.

Popham, W. James. "Defining and Enhancing Formative Assessment." Paper presented at the CCSSO State Collaborative on Assessment and Student Standards FAST meeting, Austin, Tex., October 10–13, 2006.

Pugalee, David K. "Connecting Writing to the Mathematics Curriculum." *Mathematics Teacher* 90 (1997): 308–10.

Randolph, Tamela, and Helene Sherman. "Alternative Algorithms: Increasing Options, Reducing Errors." *Teaching Children Mathematics* 7 (April 2001): 480–84.

Rohrer, Doug, and Harold Pashler. "Recent Research on Human Learning Challenges Conventional Instructional Strategies." *Educational Researcher* 39, no. 5 (2010): 406–12.

Schifter, Deborah, Stephen Monk, Susan Jo Russell, Virginia Bastable, and Darrell Earnest. "Early Algebra: What Does Understanding the Laws of Arithmetic Mean in the Elementary Grades?" In *Algebra in the Early Grades*, edited by James J. Kaput, David William Carraher, and Maria L. Blanton, pp. 413–47. New York: Lawrence Erlbaum, 2008.

Shepard, Richard G. "Writing for Conceptual Development in Mathematics." *Journal of Mathematical Behavior* 12, no. 3 (1993): 287–93.

Shulman, Lee S. "Those Who Understand: Knowledge Growth in Teaching." *Educational Researcher* 15 (1986): 4–14.

————. "Knowledge and Teaching." *Harvard Educational Review* 57, no. 1 (1987): 1–22.

Siebert, Daniel, and Nicole Gaskin. "Creating, Naming, and Justifying Fractions." *Teaching Children Mathematics* 12 (April 2006): 394–400.

Silver, Edward A., Jeremy Kilpatrick, and Beth Schlesinger. *Thinking through Mathematics: Fostering Inquiry and Communication in Mathematics Classrooms.* New York: College Board, 1990.

Steffe, Leslie P. "Children's Construction of Number Sequences and Multiplying Schemes." In *Number Concepts and Operations in the Middle Grades*, edited by James Hiebert and Merlyn Behr, pp. 119–40. Reston, Va.: National Council of Teachers of Mathematics, 1988.

————. "Children's Multiplying Schemes." In *The Development of Multiplicative Reasoning in the Learning of Mathematics*, edited by Guershon Harel and Jere Confrey, pp. vii–xxviii. Albany: State University of New York Press, 1994.

————. "Schemes of Action and Operation Involving Composite Units." *Learning and Individual Differences* 4, no. 3 (1992): 259–309.

Thompson, Patrick W., and Luis A. Saldanha. "Fractions and Multiplicative Reasoning." In *A Research Companion to "Principles and Standards for School Mathematics,"* edited by Jeremy Kilpatrick, W. Gary Martin, and Deborah Schifter, pp. 95–113. Reston, Va.: National Council of Teachers of Mathematics, 2003.

Tzur, Ron, Heather L. Johnson, Evan McClintock, Rachael H. Kenney, Yan P. Xin, Luo Si, Jerry Woodward, Casey Hord, and Xianyan Jin. "Distinguishing Schemes and Tasks in Children's Development of Multiplicative Reasoning." *PNA* 7, no. 3 (2013): 85–101.

Van de Walle, John A. *Elementary and Middle School Mathematics: Teaching Developmentally.* 6th ed. Boston: Pearson Education, 2007.

Wiliam, Dylan. "Keeping Learning on Track: Classroom Assessment and the Regulation of Learning." In *Second Handbook of Research on Mathematics Teaching and Learning*, edited by Frank K. Lester, Jr., pp. 1053–98. Charlotte, N.C.: Information Age; Reston, Va.: National Council of Teachers of Mathematics, 2007.

Yinger, Robert J. "The Conversation of Teaching: Patterns of Explanation in Mathematics Lessons." Paper presented at the meeting of the International Study Association on Teacher Thinking, Nottingham, England, May, 1998.